Mathematics Questions and Answers

P. T. Yeandle

Foreword by
Don Ewart, CEng, MIMarE, MRINA
Editor–Fairplay International Shipping Journal

STANFORD MARITIME LONDON

Stanford Maritime Limited
Member Company of the George Philip Group
12-14 Long Acre London WC2E 9LP

First published 1974
© 1974 P. T. Yeandle

Printed in Great Britain by
J. W. Arrowsmith Limited Bristol

ISBN 0 540 07337 7

Mathematics

Foreword

Written by experienced lecturers at one of Britain's leading marine engineering colleges, each book of this series is concerned with a subject in the syllabus for the examination for the Second Class Certificate of Competency. It is intended that the books should supplement the standard text books by providing engineers with numerous worked examples as well as easily understood descriptions of equipment and methods of operation. Extensive use is made of the question and answer technique and specially selected illustrations enable the reader to understand and remember important machinery details.

While the books form an important basis for pre-examination study they may also be used for revision purposes by engineers studying for the First Class Certificate of Competency.

Long experience in the operation of correspondence courses has ensured that the authors treat their subjects in a concise and simple manner suitable for individual study—an important feature for engineers studying at sea.

Don Ewart

Preface

The main object of this book is to provide students for the Second Class and First Class Engineer's Certificates of Competency with a useful reference on mathematics.

I have concentrated mainly on examination type questions and, to assist students in following the solutions, a revision maths section has been included with particular emphasis on logarithms. In the front of the book, where it is easily accessible, a list of useful formulae is given.

For convenience, the book has been divided into three parts.

1 Useful Formulae

2 Revision Maths

3 Test Papers and Solutions

PART 1

Useful Formulae

INDICES

Multiplication $\quad A^m \times A^n = A^{m+n}$

Division $\quad A^m \div A^n = A^{m-n}$

Raising the power $\quad (A^m)^n = A^{mn}$

Negative index $\quad A^{-n} = \dfrac{1}{A^n}$

Fractional index $\quad A^{1/n} = \sqrt[n]{A}$

LOGARITHMS

$$\log_x(a \times b) = \log_x a - \log_x b$$

$$\log_x(a \div b) = \log_x a - \log_x b$$

$$\log_x a^n = n \log_x a$$

$$\log_x \sqrt[n]{a} = \frac{1}{n} \log_x a$$

$$\log_x a^{-n} = -n \log_x a$$

$$\log_x x = 1$$

$$\log_x 1 = 0$$

$$\log_x a = -\log_x\left(\frac{1}{a}\right)$$

$$\log_e a = \log_{10} a \times 2 \cdot 3$$

$$e = 2 \cdot 71828$$

$$\log_e 10 = 2 \cdot 3026$$

$$\log_{10} e = 0 \cdot 4343$$

ARITHMETIC PROGRESSIONS

$$n\text{th term} = a + (n - 1)d$$

$$\therefore \quad \text{last term } (l) = a + (n - 1)d$$

1

$$S_n = \frac{n}{2}(a + l) \quad \text{when the first and last terms are known}$$

$$S_n = \frac{n}{2}\{2a + (n - 1)d\} \quad \text{when the last term is not known}$$

GEOMETRIC PROGRESSIONS

$$n\text{th term} = ar^{(n-1)}$$

$$\therefore \quad \text{last term } (l) = ar^{(n-1)}$$

$$S_n = \frac{a(1 - r^n)}{1 - r} \quad \text{to be used when } r \text{ is less than one}$$

$$S_n = \frac{a(r^n - 1)}{r - 1} \quad \text{to be used when } r \text{ is greater than one}$$

This is merely a matter of convenience to avoid negative numbers in the fraction

ALGEBRA

Simple or Linear Equations

$$y = ax + b$$

or

$$y = mx + c$$

Simultaneous Equations

If the signs of the unknown to be eliminated are the same

SUBTRACT

If the signs of the unknown to be eliminated are different

ADD

Quadratic Equations

$$y = ax^2 + bx + c$$

or

$$x = \frac{-b \pm \sqrt{b^2 - 4ac}}{2a}$$

Cubic Equations

$$x^3 + ax^2 + bx + c = 0$$

Factors

$$(a + b)^2 = a^2 + 2ab + b^2$$

$$(a - b)^2 = a^2 - 2ab + b^2$$

$$(a + b)^3 = a^3 + 3a^2b + 3ab^2 + b^3$$

$$(a - b)^3 = a^3 - 3a^2b + 3ab^2 - b^3$$

$$a^2 - b^2 = (a + b)(a - b)$$
$$a^3 + b^3 = (a + b)(a^2 - ab + b^2)$$
$$a^3 - b^3 = (a - b)(a^2 + ab + b^2)$$

TRIGONOMETRY

$$\frac{a}{\sin A} = \frac{b}{\sin B} = \frac{c}{\sin C}$$

Cosine Rule

$$a^2 = b^2 + c^2 - 2bc \cdot \cos A$$
$$b^2 = a^2 + c^2 - 2ac \cdot \cos B$$
$$c^2 = a^2 + b^2 - 2ab \cdot \cos C$$

Trigonometrical Ratios

$$\sin \theta = \frac{\text{opposite}}{\text{hypotenuse}} \qquad \cos \theta = \frac{\text{adjacent}}{\text{hypotenuse}} \qquad \tan \theta = \frac{\text{opposite}}{\text{adjacent}}$$

$$\operatorname{cosec} \theta = \frac{\text{hypotenuse}}{\text{opposite}} \qquad \sec \theta = \frac{\text{hypotenuse}}{\text{adjacent}} \qquad \cot \theta = \frac{\text{adjacent}}{\text{opposite}}$$

Ratio	Complement	Reciprocal
sine	cosine	cosecant
cosine	sine	secant
secant	cosecant	cosine
cosecant	secant	sine
tangent	cotangent	cotangent
cotangent	tangent	tangent

Basic Identities

$\sin^2 \theta + \cos^2 \theta = 1$	$\operatorname{cosec}^2 \theta - \cot^2 \theta = 1$	$\sec^2 \theta - \tan^2 \theta = 1$
$1 - \sin^2 \theta = \cos^2 \theta$	$\operatorname{cosec}^2 \theta - 1 = \cot^2 \theta$	$\sec^2 \theta - 1 = \tan^2 \theta$
$1 - \cos^2 \theta = \sin^2 \theta$	$1 + \cot^2 \theta = \operatorname{cosec}^2 \theta$	$\tan^2 \theta + 1 = \sec^2 \theta$

Compound Angles

$$\sin(A + B) = \sin A \cdot \cos B + \sin B \cdot \cos A$$
$$\sin(A - B) = \sin A \cdot \cos B - \sin B \cdot \cos A$$
$$\cos(A + B) = \cos A \cdot \cos B - \sin A \cdot \sin B$$
$$\cos(A - B) = \cos A \cdot \cos B + \sin A \cdot \sin B$$
$$\tan(A + B) - \frac{\tan A + \tan B}{1 - \tan A \cdot \tan B}$$
$$\tan(A - B) - \frac{\tan A - \tan B}{1 + \tan A \cdot \tan B}$$

3

Double Angles

$$\sin 2A = 2 \sin A \cdot \cos A$$

$$\cos 2A = \cos^2 A - \sin^2 A$$

$$= 2 \cos^2 A - 1$$

$$= 1 - 2 \sin^2 A$$

$$\tan 2A = \frac{2 \tan A}{1 - \tan^2 A}$$

$$\sin 3A = 3 \sin A - 4 \sin^3 A$$

$$\cos 3A = 4 \cos^3 A - 3 \cos A$$

$$\tan 3A = \frac{3 \tan A - \tan^3 A}{1 - 3 \tan^2 A}$$

$$\text{radians} = \theta° \times \frac{180°}{\pi}$$

Half Angles

$$\sin \tfrac{1}{2}\theta = \pm \sqrt{\frac{1 - \cos \theta}{2}}$$

$$\cos \tfrac{1}{2}\theta = \pm \sqrt{\frac{1 + \cos \theta}{2}}$$

$$\tan \tfrac{1}{2}\theta = \frac{\sin \theta}{1 + \cos \theta}$$

$$= \frac{1 - \cos \theta}{\sin \theta}$$

$$= \pm \sqrt{\frac{1 - \cos \theta}{1 + \cos \theta}}$$

$$\theta° = \text{radians} \times \frac{\pi}{180°}$$

Simpson's 1st Rule

$$= \frac{h}{3}[1 + 4 + 1] \quad \text{With minimum 3 ordinates}$$

Simpson's 1st Rule

$$= \frac{h}{3}[1 + 4 + 2 + 4 + 1] \text{ etc.} \quad \text{With more than 3 ordinates}$$

Simpson's 2nd Rule

$$= \tfrac{3}{8}h[1 + 3 + 3 + 1] \quad \text{With minimum 4 ordinates}$$

Simpson's 2nd Rule

$$= \tfrac{3}{8}h[1 + 3 + 3 + 2 + 3 + 3 + 1] \text{ etc.} \quad \text{With more than 4 ordinates}$$

THEOREM OF PAPPUS OR GULDINUS

Area swept out = length of line × distance moved by c. of g.

Volume swept out = area × distance moved by centroid.

SIMILAR FIGURES

(Ratio of linear dimensions) = n

(Ratio of areas) = (Ratio of linear dimensions)^2 = n^2

(Ratio of volumes) = (Ratio of linear dimensions)^3 = n^3

MENSURATION

Figure	Area	Volume
Square	L^2	
Cube	$6L^2$	L^3
Rectangle	$L \times B$	
Rectangular block	$2[(L \times B) + (L \times D) + (B \times D)]$	$L \times B \times D$
Parallelogram	Base $\times$ Perp Ht	
Triangle	$\dfrac{B \times H}{2}$	
	$\frac{1}{2}ab \sin C$	
	$\frac{1}{2}bc \sin A$	
	$\frac{1}{2}ac \sin B$	
	$\sqrt{s(s-a)(s-b)(s-c)}$	

where

$$s = \frac{a + b + c}{2}$$

Trapezium	$\dfrac{(a + b)}{2} \times h$	
Circle	πr^2	
	$\dfrac{\pi}{4}d^2$	
Annulus or ring	$\dfrac{\pi}{4}(D^2 - d^2)$	area $\times t$
Circle (circumference)	$2\pi r$	
	πd	
Ellipse	$\pi R r$	area $\times t$
	$\dfrac{\pi}{4}Dd$	
Ellipse (circumference)	$\pi\sqrt{\dfrac{R^2 + r^2}{2}}$	
Cone (curved surface)	$\pi r l$	
Sector of circle	$\dfrac{\pi r^2 \theta}{360°}$	
Segment of circle	$\dfrac{\pi r^2 \theta}{360°} - \dfrac{r^2 \sin \theta}{2}$	

Sphere	$4\pi r^2$	$\frac{4}{3}\pi r^3$
	πd^2	$\frac{\pi}{6}d^3$
Cylinder (solid)		$\frac{\pi}{4}d^2 l$
Cylinder (hollow)		$\frac{\pi}{4}l(D^2 - d^2)$
Cone		$\frac{\pi}{12}hd^2$ or $\frac{1}{3}\pi r^2 h$
Frustrum of cone		$\frac{\pi}{3}h(R^2 + Rr + r^2)$
		$\frac{\pi}{12}h(D^2 + Dd + d^2)$
Pyramid		$\frac{blh}{3}$
Frustrum of pyramid		$\frac{h}{3}(A + \sqrt{Aa} + a)$
Segment of sphere	$2\pi rh$	$\pi h^2\left(r - \frac{h}{3}\right)$
		$\frac{\pi}{6}h(3r_1^2 + h^2)$
		where r_1 is radius of base of segment
Sector of sphere		$\frac{2}{3}\pi r^2 h$
Sphere having hole bored through centre $\Big\}$		$\frac{4}{3}\pi(R^2 - r^2)^{\frac{3}{2}}$
Zone		$\frac{\pi k}{6}\{3(r_1^2 + r_2^2) + k^2\}$

PART 2

Revision Maths

INDICES

Definition: An index indicates how many times the base occurs as a factor of the number that is being dealt with.

$$\text{Base} \longrightarrow 3^{\overset{\displaystyle\nearrow \text{Index}}{2}} = 9 \longleftarrow \text{Number}$$

Alternatively,

$$\log_3 9 = 2 \leftarrow \text{Logarithm}$$

$$\underset{\text{Base} \quad \text{Number}}{}$$

As can be seen, the 2 has two names. Firstly it is called the index, because it indicates how many times the 3 occurs as a factor of 9.

$$3 \times 3 = 9$$

Two three's multiplied together equals nine, or $3^2 = 9$. Secondly, it is called the logarithm.

Definition: A logarithm indicates to what power the base has to be raised to give a certain number.

$$\text{Base} \longrightarrow 10^{\overset{\displaystyle\nearrow \text{Index}}{2}} = 100 \longleftarrow \text{Number}$$

In words, ten has to be raised to the power of two to give one hundred or

$$10 \times 10 = 100$$

Alternatively,

$$\log_{10} 100 = 2 \leftarrow \text{logarithm}$$

Now start right from the beginning and develop your own log tables to the base 2. Use 2 because the calculations can be done mentally, without any great effort.

Suppose a compound interest table is made up, but for simplicity an exceptional rate of interest is used—100%.

Number of years	0	1	2	3	4	5	6	7	8	9	10
Money Invested	1	1	2	4	8	16	32	64	128	256	512
Money + Interest	1	2	4	8	16	32	64	128	256	512	1024
Using base 2	2^0	2^1	2^2	2^3	2^4	2^5	2^6	2^7	2^8	2^9	2^{10}

7

It is of interest to note that the numbers in the top row (number of years) indicate the number of times that 2 occurs as a factor of the numbers in the bottom row (Money + Interest). It would not be very convenient to write:

$$2 \times 2 \times 2 \times 2 \times 2 \times 2 \times 2 \times 2 \times 2 = 512$$

It is much easier to write $2^9 = 512$, as mentioned earlier. After completing a similar table, Archimedes found that to multiply two numbers together, all that one had to do was to add the two powers from the top row together and then look up the answer in the bottom row.

For example: Multiply 8×128 which by ordinary arithmetic $= 1024$; however, using indices it can be seen that 8×128 is the same as $2^3 \times 2^7$ which is the same as $2^{3+7} = 2^{10}$ $(2 \times 2 \times 2) \times (2 \times 2 \times 2 \times 2 \times 2 \times 2 \times 2)$ which is the same as multiplying ten 2's together, which of course equals 2^{10}. Therefore looking under 10 in the top row will give the answer 1024 in the bottom row.

It is obvious that we no longer need the middle row of the table; therefore it can be rewritten using only the top row renamed as logarithms, and the bottom row renamed as antilogarithms. (Napier is generally credited with the invention of logarithms.)

It is of interest to note that the top row is in arithmetical progression, and the bottom row in geometrical progression. See page 18.

Logarithms	0	1	2	3	4	5	6	7	8	9	10
Anti-logs	1	2	4	8	16	32	64	128	256	512	1024

Now the anti-logs to base 2, can be replaced with numbers to any base we care to choose. Our common log tables are to base 10, as follows:

Logarithms	0	1	2	3	4	5	6	7	8	9	10
Anti-logs	10^0	10^1	10^2	10^3	10^4	10^5	10^6	10^7	10^8	10^9	10^{10}

$$\frac{100}{100} = \frac{10^2}{10^2} = 1 = 10^{2-2} = 10^0$$

Therefore we can say that *any* base to the power of zero is equal to one.

Remember:

To multiply—ADD the indices

To divide—SUBTRACT the indices

When indices go from a bottom line to the top line, or when they go from the top to the bottom, they change their sign; irrespective of whether it was plus or minus to start with

Example:

$$\frac{1}{10} \text{ is the same as } \frac{1}{10^1}$$

$$\frac{1}{10^1} = 0 \cdot 1 \quad \text{Which is the same as } 10^{-1}$$

If a base is raised to a negative power, it means that we have to *divide one* by the base when the same power is made positive, as shown above.

If the power is already positive, then *multiply* by one.

$$10^2 = 100 \quad \text{and} \quad 100 \times 1 = 100$$

$$10^{-2} = \frac{1}{10^2} = \frac{1}{100} = 0 \cdot 01$$

To make 10^{-2} positive it is placed under one. Suppose we had $\frac{1}{10^{-2}}$ then we would put the 10^{-2} on the top line to make it positive, as follows:

$$\frac{1}{10^{-2}} = 10^2 = 100 \quad \text{or} \quad 100 \times 1 = 100$$

To go a step further,

$$(10^2)^3 = (10^2) \times (10^2) \times (10^2) = 10^{2 \times 3} = 10^6 = 10 \times 10 \times 10 \times 10 \times 10 \times 10$$
$$= 1\,000\,000 \quad \text{or} \quad 10^6$$

Notice that we have to *multiply* the 2 and 3 together which equals 6. This is not the same as:

$$10^2 \times 10^3 = 10^{2+3} = 10^5 = 100\,000$$

where we have

$$(10 \times 10) \times (10 \times 10 \times 10) = 100\,000$$

In this case we have *added* the 2 and 3 which equals 5. Note that indices can only be added to other indices having the same base.

$$10^2 \times 10^4 \times 2^3 = 10^{2+4} \times 2^3 = 10^6 \times 2^3 = 1\,000\,000 \times 8 = 8\,000\,000$$

For the Laws of Indices see page 1.

LOGARITHMS

The Laws of Indices show us that we can express any number as the power of some other number. In other words, a logarithm tells us to what power the base has to be raised to give a certain number. See pages 7 and 12.

So far however, we have only dealt with whole numbers. The whole number part of a logarithm is called the CHARACTERISTIC. More often than not, logarithms are made up of a whole number and a decimal part. The decimal part is called the MANTISSA, and it is the mantissa that is tabulated in our tables of Common Logarithms.

It is extremely important that students realize right from the very beginning that the values shown in our log tables are POSITIVE values. The importance of this statement will not be appreciated until we deal with negative characteristics later on.

Example:

$$10^2 = 100 \quad \text{or} \quad \log_{10} 100 = 2 \cdot 0000$$

Characteristic   Mantissa

Logarithm

$$10 \times 10 = 100 \quad \text{or} \quad \log 100 = 2 \cdot 0000$$

Note that we do not normally indicate the base when using a log to the base ten. Therefore, if no base is shown we take it for granted that it *is* ten.

$$10 \times 10 \times 10 = 1000 \quad \text{or} \quad \log 1000 = 3.0000$$

What would we write if we had $10^{2.5}$? $10^{2.5} = 10 \times 10 \times$? we know that we have two tens because the characteristic tells us so; but what does the $\cdot5$ tell us? We could write $10^{2.5}$ like this; $10^1 \times 10^1 \times 10^{\frac{1}{2}}$ which would be the same as $10^{1+1+\frac{1}{2}} = 10^{2\frac{1}{2}}$. When using logarithms, never lose sight of the fact that we are dealing with powers and roots.

Remember: *A tree*—powers at the top and roots at the bottom *like the roots of a tree*. Therefore $10^{\frac{1}{2}}$ could be written as $\sqrt[2]{10^1}$, but as you know, we do not normally write in the power when it is only one, nor do we write in the root if it is two; so the usual way to write $\sqrt[2]{10^1}$ would be $\sqrt{10}$ which of course is equal to 3·162. So, $10^{2.5} = 10 \times 10 \times \sqrt{10} = 10 \times 10 \times 3.162 = 316.2$ alternatively, $\log 316.2 = 2.5000$. When using common logs we only take the decimal part (mantissa) from the tables, the characteristic is decided by us. The rule is: that the characteristic is one less than the number of figures to the left of the decimal point. See below.

$$1000 = 3.0000$$
$$100 = 2.0000$$
$$10 = 1.0000$$
$$1 = 0.0000$$
$$0.1 = \bar{1}.0000$$
$$0.01 = \bar{2}.0000$$
$$0.001 = \bar{3}.0000$$

When we use negative characteristics, the number is the same as the number of noughts. For example 0·001; three noughts, therefore $\bar{3}$, or 0·000001; six noughts, $\bar{6}$. The minus sign over the top of a characteristic indicates that only the three is negative. $\bar{3}.4784$ is the same as $-3 + 0.4784$; the 3 is negative, the ·4784 positive.

$$3.4784 \quad \text{all positive}$$
$$\bar{3}.4784 \quad \text{part positive, part negative}$$
$$-3.4784 \quad \text{all negative}$$

A part positive, part negative number can be written as a wholly negative number.

$$\bar{3}.4784 \text{ could be written as } -2.5216$$

To do this, take the difference between -3 and $+0.4784$

$$-3.0000$$
$$+0.4784$$
$$\text{Difference} \quad -2.5216$$

$\bar{3}.4784$ has exactly the same value as -2.5216, it is only written in a different form. Suppose we take

$$\log 0.0272 = \bar{2}.4346$$
$$= -2 + 0.4346$$
$$\text{Difference} = -1.5654$$

This can be quite easily understood if we draw a thermometer type diagram (see Fig. 2.1).

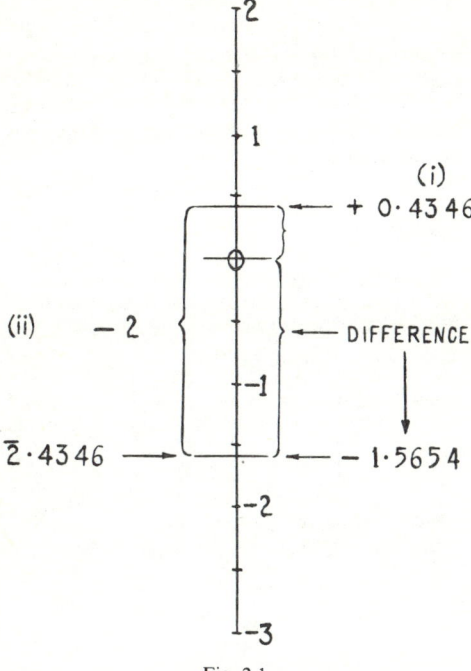

Fig. 2.1

(i)   Mark off +0·4346
(ii) Now mark off −2 downwards from (i) above.

Up for positive values, down for negative.

It must be clearly understood that −1·5654 is the *difference* between −2 and +0·4346. Therefore when we write −1·5654 it is just a different way of writing $\bar{2}$·4346, because they *both* represent exactly the same value.

The fact that we can change a part positive and part negative logarithm into a wholly negative value is extremely useful because it simplifies the raising of values less than one to various powers, which will be explained later.

Once we have a wholly negative value, it is necessary to find a way of converting it back into positive form for the mantissa. This is something that we have to do in order to be able to use the log tables, as the tabulated values are all positive. The characteristic remains negative.

Referring once again to −1·5654, we are now going to convert it back to $\bar{2}$·4346. This is accomplished by adding +1 to the decimal part, and −1 to the whole number.

$$-1 \quad +1{\cdot}0000$$

$$\frac{-1 \quad -0{\cdot}5654}{-2 \quad +0{\cdot}4346} \quad \text{and putting them together, gives us}$$

$$\bar{2}{\cdot}4346$$

11

Remember that we have to put the minus sign over the top of the 2 now, to indicate that only the 2 is negative.

We always add $+1$ to the side that has to be positive in order to use the tables i.e., the decimal part (mantissa), and it is *always* $+1$ that we *add*, never any other number.

By adding -1 on one side and $+1$ on the other, the total effect, mathematically, is zero. So if we have added *nothing* to a certain quantity, we cannot have changed its value. We have only succeeded in changing its form, or the way we write it; which, of course, is exactly what we wanted to do.

Now do some examples.

1)   Multiply 28·62 by 9·38

No.	Logarithm	
28·62	1·4567	(Remember to use the difference column)
9·38	0·9722	

To multiply *add* logs   2·4289

Antilogging = 268·5   *Answer*

2)   Divide 465·34 by 29·2

No.	Log	(We usually abbreviate the word logarithm to
465·34	2·6678	log)
29·20	1·4654	

To divide *subtract*   1·2024

Antilogging = 15·93   *Answer*

3)   To give a better understanding of what we mean, when we say that a logarithm tells us to what power the base has to be raised to give a certain number; do the same problem again using indices. See page 7.

Divide 465·34 by 29·2

$$465.34 = 10^{2.6678}$$

$$29.20 = 10^{1.4654}$$

$$\therefore \quad 465.34 \div 29.2 = 10^{2.6678} \div 10^{1.4654}$$

$$= 10^{2.6678 - 1.4654}$$

$$= 10^{1.2024}$$

Antilogging = 15·93   *Answer*

Raising the Power

4)   Find the value of $w = 27 \cdot 2^2$, using logarithms:

$$w = 2 \times \log 27 \cdot 2$$

$$= 2 \times 1.4346$$

$$= 2.8692$$

Antilogging $w = 939 \cdot 1$ *Answer*

5) Now move the decimal point one place, but still using the same figures find the value of:

$$2 \cdot 72^2 = 2 \times \log 2 \cdot 72$$
$$= 2 \times 0 \cdot 4346$$
$$= 0 \cdot 8692$$

Antilogging $= 9 \cdot 391$ *Answer*

6) In a similar manner and moving the decimal point another place:

$$0 \cdot 272^2 = 2 \times \log 0 \cdot 272$$
$$= 2 \times \bar{1} \cdot 4346$$
$$= \bar{2} \cdot 8692$$

Antilogging $= 0 \cdot 09391$ *Answer*

7) So far we have only raised several numbers to the power of 2, now we will raise 0·272 to the power of 2·46:

$$0 \cdot 272^{2 \cdot 46} = 2 \cdot 46 \times \log 0 \cdot 272$$
$$= 2 \cdot 46 \times \bar{1} \cdot 4346$$

At this stage you will see that the multiplication is not as straightforward as the previous example because, instead of multiplying by one figure (2) we now have to multiply by three figures (2·46). To simplify the multiplication it is necessary to convert $\bar{1} \cdot 4346$ to a wholly negative number. See page 10.

By taking the difference between -1 and $+0 \cdot 4346$ we will get $-0 \cdot 5654$, which of course has exactly the same value as $\bar{1} \cdot 4346$.

To continue

$$2 \cdot 46 \times \bar{1} \cdot 4346 = 2 \cdot 46 \times -0 \cdot 5654$$
$$= -1 \cdot 390884$$

But using four figure logs we call it:

$$= -1 \cdot 3909$$

We now have to convert the wholly negative number back into part positive, part negative form. See page 11.

$$-1 + 1$$
$$= -1 \cdot 3909$$
$$= \bar{2} \cdot 6091$$

Antilogging $= 0 \cdot 04065$ *Answer*

Do not forget that when multiplying two values together they take the algebraic result of the signs, as follows:

$$+ \times + = +$$
$$- \times - = +$$
$$+ \times - = -$$

The object being to get positive values in order to use the log tables.

LOGLOG is an abbreviation for the logarithm of a logarithm. Sometimes, when multiplying a log by a power it is easier to use logs rather than do the operation by arithmetic, which can be very tedious if you have a lot of decimals. Therefore repeat the previous problem using loglogs.

8) $$0.272^{2.46} = 2.46 \times \log 0.272$$

$$= 2.46 \times \bar{1}.4346$$

$$= 2.46 \times -0.5654 \quad \text{(At this point we can use loglogs)}$$

log of the log of 0.272; is the same as, the log of $\bar{1}.4346$ but to take the log of $\bar{1}.4346$ it is necessary to make it into a wholly negative number, as shown above:

$$\log \quad 2.46 \quad = 0.3909 \quad \text{(When logging } -0.5654; \text{ treat as a } \textit{positive} \text{ value)}$$
$$\log -0.5654 = \underline{1.7523}$$
$$0.1432$$

Now we do our first antilogging = 1.3910.

Now apply the algebraic result of the signs: $+ \times - = - = -1.3910$. Before we can do our second antilogging, convert to a part positive, part negative number by adding -1 and $+1$.

$$-1 + 1$$
$$-1.3910$$
$$= \overline{2}.6090$$

Now we can do our second antilogging

$$= 0.04064 \quad \textit{Answer}$$

Note that there is a very slight difference of 0.00001 in the answer. This is because we have used four figure logs which are not quite as accurate as arithmetic.

Once again using the same values, but this time make the power negative.

9) $$0.272^{-2.46} = -2.46 \times \log 0.272$$

$$= -2.46 \times \bar{1}.4346$$

$$= -2.46 \times -0.5654 \quad \text{(Using loglogs)}$$

$$\log -2.46 = 0.3909$$
$$\log -0.5654 = \underline{\bar{1}.7523}$$
$$0.1432$$

So far this *looks* the same as the previous example, but it is not quite the same. From our first antilogging we get 1.3910. Now apply the algebraic result of the signs $- \times - = +$. As the result is positive, it is not necessary to make the mantissa positive; we only have to do the second antilogging.

$$\text{Second antilogging} = 24.60 \quad \textit{Answer}$$

10) A further example:

$$27.2^{-2.46} = 2.46 \times \log 27.2$$

or

$$\frac{1}{27 \cdot 2^{2 \cdot 46}} = \log 1 - 2 \cdot 46 \times \log 27 \cdot 2$$

$$= 0 \cdot 0000 - 2 \cdot 46 \times 1 \cdot 4346$$

$$= 0 \cdot 0000 - 3 \cdot 5291$$

$$= \bar{4} \cdot 4709$$

Antilogging $= 0 \cdot 0002957$ *Answer*

Note that we used arithmetic in this instance, although loglogs could have been used should we wish to do so. Also your attention is directed to the fact that we have made the power positive by placing both the number and power, under one. The advantage of using this method, is that we avoid a wholly negative number and then having to add -1 and $+1$, to make the mantissa positive. As an exercise, perhaps you would like to do this and see if you get the same answer.

ROOTS OF NUMBERS

Remember: Roots at the bottom, like the roots of a tree.

11) $$4^{\frac{1}{2}} = \frac{1}{2} \times \log 4$$

$$= \frac{1}{2} \times 0 \cdot 6021 = \frac{0 \cdot 6021}{2} = 0 \cdot 3010$$

Antilogging $= 2$ *Answer*

12) $$0 \cdot 4^{\frac{1}{2}} = \frac{1}{2} \times \log 0 \cdot 4$$

$$= \frac{1}{2} \times \bar{1} \cdot 6021 = \frac{\bar{1} \cdot 6021}{2} = \frac{\bar{2} + 1 \cdot 6021}{2}$$

$$= \bar{1} + 0 \cdot 8010 = \bar{1} \cdot 8010$$

Antilogging $= 0 \cdot 6324$ *Answer*

To achieve this result, it is necessary to add whatever *negative* value may be required to the negative characteristic, to enable us to divide the root into it without leaving a remainder. Secondly, whatever negative value has been added to the characteristic must now be added in *positive* form to the mantissa.

13)   Find the value of

$$0 \cdot 00867^{\frac{3}{4}} = \frac{3}{4} \times \log 0 \cdot 00867$$

$$= \frac{3}{4} \times \bar{3} \cdot 9380$$

$$= \frac{3 \times \bar{3} \cdot 9380}{4} = \frac{\bar{7} \cdot 8140}{4}$$

$$= \frac{\bar{8} + 1 \cdot 8140}{4}$$

$$= \bar{2} + 0 \cdot 4528$$

$$= \bar{2} \cdot 4528$$

Antilogging $= 0 \cdot 02836$ *Answer*

14) Find the value of

$$\sqrt{\sqrt[4]{0\cdot638}} = 0\cdot638^{\frac{1}{8}}$$

(Multiply the roots together in this instance)

$$= \frac{\log 0\cdot638}{8} \qquad (\tfrac{1}{2} \times \tfrac{1}{4} = \tfrac{1}{8})$$

$$= \frac{\bar{1}\cdot8048}{8}$$

$$= \frac{\bar{8} + 7\cdot8048}{8}$$

$$= \bar{1} + 0\cdot9756$$

$$= \bar{1}\cdot9756$$

Antilogging $= 0\cdot9454$ *Answer*

Finding the value of an unknown power

15) $1\cdot8^n = 8\cdot306$, find the value of 'n':

$$n \log 1\cdot8 = \log 8\cdot306$$

$$\therefore \quad n = \frac{\log 8\cdot306}{\log 1\cdot8} = \frac{0\cdot9194}{0\cdot2553} = 3\cdot6 \quad \text{(Do not antilog)}$$

$$n = 3\cdot6 \quad \textit{Answer}$$

Note that we *divide* the log of $8\cdot306$ by the log of $1\cdot8$ by *ordinary arithmetic*; you do NOT subtract the logs. If you wish to use logs then you must use loglogs and antilog ONCE.

16)   If $0\cdot75^n = 0\cdot5$; find the value of 'n'

$$n \times \log 0\cdot75 = \log 0\cdot5$$

$$n = \frac{\log 0\cdot50}{\log 0\cdot75}$$

$$n = \frac{\bar{1}\cdot6990}{\bar{1}\cdot8751} = \frac{-0\cdot3010}{-0\cdot1249} = 2\cdot4098$$

$$n = 2\cdot4098 \quad \textit{Answer}$$

17)   Evaluate for 'x' in:

$$5x = (9\cdot6)^{x-3}$$

$$x \, . \log 5 = (x - 3) \, . \log 9\cdot6$$

$$0\cdot6990x = 0\cdot9823(x - 3)$$

$$0\cdot6990x = 0\cdot9823x - 2\cdot9469$$

$$2\cdot9469 = 0\cdot9823x - 0\cdot6990x$$

$$2\cdot9469 = 0\cdot2833x$$

$$\therefore \quad x = \frac{2\cdot9469}{0\cdot2833} = 10\cdot4$$

$$x = 10\cdot4 \quad \textit{Answer}$$

NAPIERIAN LOGARITHMS

Natural or Napierian logarithms are to the base 'e', which is equal approximately to 2·7183. Multiplication, division, raising powers, taking roots, etc., are dealt with in exactly the same way as logs to base 10 (common logs).

The main differences are:

a)   The characteristic is given in the tables.

b)   Using the Napierian log tables.

c)   If the number is less than one or more than 9·999, then for every log to base 10, it will be necessary to use two logs to base 'e'.

Example:

18)   Multiply 29·7 × 8·64 × 0·035

	Using logs to base 10		Using logs to base 'e'			
No.	Logs$_{10}$	No.	Logs$_e$	No.	Logs$_e$
29·7	1·4728	29·7	3·3912	2·97	1·0886
8·64	0·9365	8·64	2·1564	10^1	2·3026
0·035	$\overline{2}$·5441	0·035	$\overline{4}$·6476	29·7	3·3912
	0·9534		2·1952	3·5	1·2528
				10^{-2}	$\overline{3}$·3948
				0·035	$\overline{4}$·6476

Antilogging = 8·982   *Answer*   Antilogging = 8·982   *Answer*

As can be seen from the above example, for values below 1 and above 9·999 it is necessary to use the small tables at the bottom of the page. The small tables give logs$_e$ for various powers of 10, both positive (on the left hand side) and negative on the right. Therefore, for values below 1 it is the log$_e$ of a number taken from the main table (reference 3·5 above), added to the log$_e$ of whatever power 10 is raised to (in this case 10^{-2}), to obtain the log$_e$ of the number required (0·035). The log$_e$ 10^{-2} being taken from the table at the foot of the right hand page, because it is a *negative* power. For numbers above 9·999 a similar procedure is adopted, but in this case use the small table at the foot of the left hand page because we then use *positive* powers.

It should be noted that 8·64 was a direct reading from the main table.

It is a useful exercise to work out a problem using common logs, and then to work out the same problem again using Napierian logs. Exactly the same answer should be obtained from each method.

Antilogging Napierian logs

Antilog the following logs$_e$

19)   2·0968   As this log$_e$ is more than 0·0000 and less than 2·3026, it is a direct reading from the main table. Remember that the characteristic is *included* and runs horizontally across the page. Start by looking in the middle of the table and then look to the extreme left column and vertically above to the top of the page; when it should be noted that we get

     8·14   *Answer*

20) Now antilog 5·8542   Looking in the main table it is clearly seen that this number is not included; therefore as it is a positive characteristic, look in the small table at the foot of the left hand page, when it will be seen that it comes somewhere between columns 2 and 3. We now take the next *lowest number* and *subtract* it from our log. This will then leave a number with which you can enter the main table with. The number 2 at the top of the column from which the next lowest number (log) was taken, can be treated in exactly the same way as a characteristic in common logs. Therefore, there will be three figures to the left of the decimal point. The last figure (7) was obtained from the difference column, as the nearest we could get to 1·2490 was 1·2470 which left us with 0·0020 to enter the difference column with.

$$
\begin{array}{l}
5·8542 \\
6·6052 \\
\hline
1·2490 \\
\\
= 3487 \quad \textit{Answer}
\end{array}
$$

We will now antilog a log$_e$ with a negative characteristic.

21) Antilog $\bar{8}$·3769   Once again it will be noticed that this log is not included in the main table, therefore as we have a *negative* characteristic we look in the small table at the foot of the *right* hand page.

$$
\begin{array}{l}
\bar{8}·3769 \\
\overline{10}·7897 \\
\hline
1·5872 \\
\\
= 0·000489 \quad \textit{Answer}
\end{array}
$$

It will be seen that $\bar{8}$·3769 comes between columns 3 and 4. The next lowest number in this instance is $\overline{10}$·7897, because it is more negative than $\bar{8}$·3769. $\overline{10}$·7897 is now *subtracted* from $\bar{8}$·3769, we then proceed as before. The next lowest number (log) was taken from column 4, and once again this can be treated the same as a characteristic in common logs, giving us four noughts in the answer (0·000).

Should any difficulty be experienced in deciding which is the next lowest number, only the characteristics need be considered if they are different; it is then an easy matter to decide which is the most negative. In this case − 10 ($\overline{10}$) is obviously more negative than −8 ($\bar{8}$), therefore it must be the lowest number. Should the characteristics be the same, as for example in $\bar{7}$·0922 (see column 3, at foot of right hand page) and $\bar{7}$·0923, then only the mantissa from each log need be compared. Remember that we are now dealing with positive values, so it must be equally obvious that 0·0922 is lower than 0·0923. In other words, $\bar{7}$·0922 is lower, more negative, than $\bar{7}$·0923.

If difficulty is still experienced, then the logs could be converted to wholly negative numbers as already demonstrated on page 10.

$$\bar{7}·0923 = -6·9077$$
$$\bar{7}·0922 = -6·9078$$

From this it can be very clearly seen that − 6·9078 ($\bar{7}$·0922) is quite definitely lower and more negative than − 6·9077 ($\bar{7}$·0923).

PROGRESSIONS

Arithmetical Progressions are a series of terms separated by a common difference which may be either positive or negative.

Example:

$$1, \ 2, \ 3, \ 4, \ 5, \ 6, \quad \text{Common difference} = 1$$
$$24, 27, 30, 33, 36, 39, \quad \text{Common difference} = 3$$
$$42, 35, 28, 21, 14, \ 7, \quad \text{Common difference} = -7$$

where

n = the number of terms

a = the first term

d = the common difference

l = the last term

S_n = the sum of 'n' terms

Geometrical Progressions are a series of terms connected by a common ratio which may be either positive or negative. The common ratio may be found by dividing any term by its preceding term.

Example:

$$1, \quad 3, 9, \quad 27, 81, \quad 243, \quad \text{Common ratio} = 3$$
$$\tfrac{1}{4}, \quad \tfrac{1}{2}, 1, \quad 2, \ 4, \quad 8, \quad \text{Common ratio} = 2$$
$$10, -5, \tfrac{5}{2}, -\tfrac{5}{4}, \tfrac{5}{8}, -\tfrac{5}{16}, \quad \text{Common ratio} = -\tfrac{1}{2}$$

where

n = the number of terms

a = the first term

r = the common ratio

S_n = the sum of 'n' terms

ALGEBRA

Quantities in algebra are treated in a similar manner to those in arithmetic, but in a more general sense. In arithmetic a quantity means exactly what it says, it has one definite value. In algebra there is a greater freedom, and quantities are denoted by symbols—usually letters of our own alphabet—which may represent any value we wish to give them.

In solving problems it is important that students are able to recognize the sort of equation they may produce in the middle of a solution.

For example $(x^2 - 1)$ may appear, which should immediately be recognized as being in the form of $(a^2 - b^2)$ which is equal to $(a + b)(a - b)$.

Therefore $(x^2 - 1)$ could be written as $(x + 1)(x - 1)$.

These general formulae are given as *tools* to be used in helping to solve the problem. We would not bang a nail in with a screwdriver, we would use a hammer. So it is with particular equations, they each have their special methods for solution. The only way to become familiar with these methods is by constant practice.

The most important points in dealing with mathematical problems are:

a) *Interpretation*: Read the question word for word and make quite sure that you know *exactly* what is required. This is not always easy.

19

b) *Transposition*:   This means changing the basic formula into a different form.

22) Example:

$$\text{Area of a circle} = \pi r^2$$

But suppose we want a formula with which to find r, then we would have to transpose the formula into:

$$r = \sqrt{\frac{\text{Area}}{\pi}}$$

c) *Recognition*: It is important to recognize the type of equation with which you are dealing, in order to use the right method to solve it. It may be necessary to use several different formulae in one solution, and it is extremely important that you recognize each step as it appears. In other words, use the right *tools*.

Remember:

a, $1a$, a^1, $1a^1$, all mean exactly the same.

$$- \times - = +$$
$$+ \times + = +$$
$$- \times + = -$$

$$\frac{-}{-} = + \qquad \frac{+}{+} = + \qquad \frac{+}{-} = - \qquad \frac{-}{+} = -$$

$$\frac{1}{1} = 1 \qquad \frac{1}{x \to 0} \to \infty \qquad \frac{0}{1} = 0 \qquad 0^1 = 0 \qquad 1^0 = 1$$

$$a + a = 2a \qquad a + b = a + b$$
$$a - a = 0 \qquad a - b = a - b$$
$$a \times a = a^2 \qquad a \times b = ab$$
$$\frac{a}{a} = 1 \qquad \frac{a}{b} = \frac{a}{b}$$

$$\text{OR} \quad \frac{a^1}{a^1} = a^{1-1} = a^0 = 1 \qquad \text{OR} \quad \frac{a^1}{b^1} = a^1 \times b^{-1} = \frac{a}{b}$$

Addition

23) Only terms of the same order may be added; e.g.

$$7a + 5a = 12a$$
$$7a - (-5a) = 12a$$
$$7a + (-5a) = 2a$$
$$7a - 5a = 2a$$

$$7a + 5b = 7a + 5b$$
$$7a - (-5b) = 7a + 5b$$
$$7a + (-5b) = 7a - 5b$$
$$7a - 5b = 7a - 5b$$

Find the sum of means the same as *add together*; e.g.

24)
$$4x + 3y - 3z$$
$$-3x + 2y + 2z$$
$$\underline{2x - 4y + 4z}$$
$$3x + \ \ y + 3z$$

25)
$$19ab - 23bc - \ \ 2ca$$
$$18ab - 14bc + 14ca$$
$$\underline{-45ab + 49bc - \ \ 3ca}$$
$$-8ab + 12bc + \ \ 9ca$$

26)
$$x^3 - 6x^2y + \ \ 8xyz$$
$$2x^3 + 2x^2y - 12xyz$$
$$\underline{-2x^3 + 3x^2y + \ \ xyz}$$
$$x^3 - \ \ x^2y - \ \ 3xyz$$

27)
$$\tfrac{1}{2}x - \tfrac{2}{3}y$$
$$-x + \tfrac{1}{3}y$$
$$\underline{\tfrac{5}{4}x - \tfrac{5}{6}y}$$
$$\tfrac{3}{4}x - \tfrac{7}{6}y$$

28)
$$\tfrac{7}{8}a^2 - \tfrac{1}{3}ab + \tfrac{3}{10}b^2$$
$$-\tfrac{5}{4}a^2 + \tfrac{13}{15}ab - \ \ b^2$$
$$\underline{\tfrac{1}{2}a^2 - \ \ ab + \tfrac{1}{5}b^2}$$
$$\tfrac{1}{8}a^2 - \tfrac{7}{15}ab - \tfrac{1}{2}b^2$$

Subtraction

In a similar manner to addition, only terms of the same order (power or degree) can be subtracted from one another.

<u>Rule:</u>   Change the sign of the *bottom* line and *add*.

29) e.g.

$$
\begin{array}{ll}
\text{From} & 14a + \ \ 9b - 17c \\
\text{Subtract} & \underline{\ \ 2a - \ \ 8b + \ \ c} \\
& 12a + 17b - 18c
\end{array}
$$

Subtract $9x + 7y - 6z$ from $3x - 2y + 11z$.

30)
$$3x - 2y + 11z$$
$$\underline{9x + 7y - \ \ 6z}$$
$$-6x - 9y + 17z$$

31)
$$\tfrac{3}{4}a + 2b - 2c$$
$$\underline{\tfrac{1}{2}a - \tfrac{3}{2}b - \tfrac{4}{3}c}$$
$$\tfrac{1}{4}a + \tfrac{7}{2}b - \tfrac{2}{3}c$$

Multiplication

In its simplest form multiplication really means repeated addition.

$$\text{e.g.} \quad 5 \times 6 = 30$$

OR
$$5 + 5 + 5 + 5 + 5 + 5 = 30 \quad \text{(5 taken 6 times)}$$

OR $\qquad$ $6 + 6 + 6 + 6 + 6 = 30$ (6 taken 5 times)

$\qquad$ $a + a + a + a = 4a$ or $4 \times a$

$\qquad$ $\frac{3}{4} \times \frac{1}{2} = \frac{\frac{3}{8} + \frac{3}{8}}{\frac{2}{1}} = \frac{\frac{2}{1} \times \frac{3}{8}}{\frac{2}{1}} = \frac{\frac{6}{8}}{\frac{2}{1}} = \frac{3}{4} \times \frac{1}{2} = \frac{3}{8}$

OR $\qquad$ $\frac{3}{4} \times \frac{1}{2} = \frac{3}{16} + \frac{3}{16} = 2(\frac{3}{16}) = \frac{3}{8}$

To multiply numbers raised to a power we *multiply* the coefficients and *add* the indices.

32) $\qquad$ $4x^2 \times 8x^5 = 4 \times 8 \times x^{2+5} = 32x^7$ *Answer*

33) $\qquad$ $7x^3y^2z^5 \times 11xyz^9 = 77x^4y^3z^{14}$ *Answer*

34) $\qquad$ $(xy + yz) \times x^5z^2 = x^6yz^2 + x^5yz^3 = x^5yz^2(x + z)$ *Answer*

Note in this last example, the highest common factor (HCF) that will go into each term has been taken outside the bracket.

35) $\qquad$ $xy + yz \times x^5z^2 = xy + x^5yz^3 = xy(1 + x^4z^3)$ *Answer*

<u>Note</u>:   If there is no bracket, multiplication *always* takes priority.

Division

To divide terms raised to a power, divide the coefficient and subtract the indices.

In long division where there is more than one term in the divisor, adopt the following procedure:

(a) Put the terms of both the divisor (terms we are going to divide by) and the dividend (equation we are going to divide into) in descending order (powers) of some common letter.

(b) Divide the *first* term of the divisor into the *first* term of the dividend, this will give the quotient (answer).

(c) Multiply *every term* in the divisor by the term that has last been added to the quotient.

(d) Subtract by changing the sign of the bottom line and adding.

(e) The first term will always cancel out, but if any of the following terms cannot be subtracted then change the sign and place it underneath with its new sign, and bring down the term above it but DO NOT change the sign of any term brought down from the top line. Remember to place the terms in descending order.

(f) Start all over again, always the first term into the first term.

36) Example:

$$2x + 9)\overline{12x^4 + 62x^3 + 26x^2 - 27x + 81}(6x^3 + 4x^2 - 5x + 9$$

$$
\begin{array}{r}
12x^4 + 54x^3 \\
\hline
8x^3 + 26x^2 \\
8x^3 + 36x^2 \\
\hline
-10x^2 - 27x \\
-10x^2 - 45x \\
\hline
18x + 81 \\
18x + 81 \\
\hline
\qquad
\end{array}
$$

Answer $6x^3 + 4x^2 - 5x + 9$

22

37)
$$\frac{a^3 + b^3}{a + b} = \quad a + b)\overline{a^3 + b^3}(a^2 - ab + b^2$$
$$\underline{a^3 + a^2b}$$
$$-a^2b + b^3$$
$$-a^2b - ab^2$$
$$\underline{}$$
$$ab^2 + b^3$$
$$ab^2 + b^3$$

Answer $\quad a^2 - ab + b^2$

ALGEBRAIC FACTORS

A FACTOR is any number which divides *exactly* into another number.

A MULTIPLE is any number which contains another number an *exact* number of times.

A COMMON FACTOR is a number which will divide *exactly* into each of two or more numbers.

Example:

$$2, 3, 4, 6, 8, 12, 16, 24 \quad \text{are all } \textit{factors of } 48$$

While

$$9, 27, 30, 33 \quad \text{are all } \textit{multiples of } 3$$

$\therefore$ 3 is also a *common factor* of the above numbers.

HIGHEST COMMON FACTOR is the expression of highest value which will divide each of two or more expressions without leaving a remainder.

38) The HCF of $a^2 - ab$ and $a^2 - b^2$ is $(a - b)$ because

$$a(a - b) = (a^2 - ab) \quad \text{and} \quad (a + b)(a - b) = (a^2 - b^2)$$

$\therefore$ HCF is $(a - b)$ *Answer*

The HCF of 7, 14, 21, 28, and 35 is 7

REMEMBER

$$(a^2 - b^2) = (a + b)(a - b)$$
$$(a - b)^2 = a^2 - 2ab + b^2$$
$$(a + b)^2 = a^2 + 2ab + b^2$$
$$(a^3 - b^3) = (a - b)(a^2 + ab + b^2)$$
$$(a^3 + b^3) = (a + b)(a^2 - ab + b^2)$$

If we multiply two brackets together and get a quadratic equation, as follows:

$$(x - 4)(x + 9) = 0$$
$$x(x + 9) - 4(x + 9) = 0$$
$$x^2 + 9x - 4x - 36 = 0$$
$$x^2 + 5x - 36 = 0$$

23

we can factorize this equation as shown below. But *remember* that all quadratic equations cannot be factorized.

To Factorize

39) $$x^2 + 5x - 36 = 0$$

Multiply x^2 by -36 to get the product (P)
The factors of $-36x^2$ when ADDED (sum) together should give the middle term $5x$. Expand the equation to four terms by substituting these two factors for the middle term $5x$.

$$\overbrace{x^2 - 4x} + \overbrace{9x - 36} = 0$$

Separate into two pairs and factorize

$$x(x - 4) + 9(x - 4) = 0$$

$$(x + 9)(x - 4) \quad = 0$$

$P = -36x^2$
$S = 5x$

1×36
2×18
3×12
4×9
6×6
9×4

Factors are $-4x$ and $9x$
which of course $= 5x$

When factorizing, it must not be forgotten that each *pair* of factors can give FOUR different answers; therefore it is extremely important to make sure that the factors, which are substituted for the middle term in the equation, not only have the correct figures but also have the correct SIGNS.

$$+9x + 4x = +13x$$

$$-9x - 4x = -13x$$

$$+9x - 4x = +5x \leftarrow \text{correct pair}$$

$$-9x + 4x = -5x$$

If any value is multiplied by zero (0), the answer must be zero. Alternatively, if two values when multiplied together *equal* zero, then either one, or *both* of these values could be equal to zero. Therefore take them one at a time as follows:

$$(x + 9)(x - 4) = 0$$

If $(x + 9) = 0$; then $x = -9$
If $(x - 4) = 0$; then $x = 4$ } *by transposition*

So the answer must be that x could be equal to either -9 or 4.

Answer $x = -9$ or 4

Perfect Square

It cannot be emphasized too often how important it is that students should be able to distinguish between the various types of equations; in this particular instance, between an ordinary quadratic equation and a perfect square. For example

$$x^2 + 10x + 21 = 0 \quad \text{is not the same sort of equation}$$

as

$$x^2 + 10x + 25 = 0 \quad \text{although it may look the same.}$$

If we factorize

$$x^2 + 10x + 21 = 0$$

we get

$$(x + 3)(x + 7) = 0$$

But if we factorize

$$x^2 + 10x + 25 = 0$$

we will get

$$(x + 5)(x + 5) = 0$$

As can be seen, the first equation has *different* values in each bracket. The second equation has the *same* values, and is known as a PERFECT SQUARE because it is the perfect square of its factors.

To establish whether the equation is a perfect square or not, we adopt the following procedure.

$$x^2 + 10x + 25 = 0$$

Take the square root of $x^2 = x$, and the square root of $25 = 5$. Multiply x by $5 = 5x$. Double this value $= 5x \times 2 = 10x$. If this value $(10x)$ is equal to the middle term—as it is in this case—then the equation is a perfect square.

When solving a quadratic equation by the COMPLETING THE SQUARE method, the whole object is to make an equation of the first type into an equation of the second type. If the equation is already a perfect square then there is no point in completing the square.

Difference of Squares

The difference of the squares of two quantities is the product of their sum and difference. i.e.,

Example:
$$(a^2 - b^2) = (a + b)(a - b) \quad \text{General Form}$$
$$(x^2 - 1) = (x + 1)(x - 1) \quad \text{because } 1^2 = 1$$

$(a^2 - b^2)$ occurs frequently in various solutions and in various forms. It is up to the student to recognize it when he sees it.

Example:
The area of a circular metal plate with a hole cut in the middle, could be given as:

$$\pi R^2 - \pi r^2 = \pi(R^2 - r^2) \quad \text{Taking out HCF}$$
$$= \pi(R + r)(R - r) \quad \text{Applying } (a^2 - b^2)$$

or $\cot^2 18° - \cot^2 20° = (\cot 18° + \cot 20°)(\cot 18° - \cot 20°)$ is an example using trigonometrical ratios.

Fractions

It is often necessary to deal with units that are not *whole* numbers. In addition to this, it is quite likely that the units involved are not of the same kind.

Example:
$\frac{1}{4}$ is different to $\frac{4}{16}$ although it may be of the same value, as it is in this case. Therefore, in

25

order to add or subtract fractions we have to use a method that will make the fractions into the *same* kind of units. e.g.,

$$\frac{1}{4} + \frac{1}{2} = \frac{1}{4} + \frac{2}{4} = \frac{1+2}{4} = \frac{3}{4}$$

To enable us to do this, find the Lowest Common Multiple (LCM); in other words find a *common* denominator, as follows:

40) $\dfrac{5}{8} + \dfrac{5}{12} + \dfrac{3}{18} + \dfrac{1}{4} + \dfrac{2}{3}$ ← Numerator
                                                                                           ← Denominator

Now write each of the denominators down and—using only prime numbers, a prime number being a number which has no factors other than itself and unity (1)—starting with 2, so long as at least one of the denominators can be divided by 2; keep dividing by this number until you find that there is nothing left that can be divided any more by 2. Then start with 3, and so on until there is nothing left to divide. All the divisors are now multiplied together to give the LCM, as shown below.

2	3,	4,	8,	12,	18
2	3,	2,	4,	6,	9
2	3,	—	2,	3,	9
3	3,	—	—	3,	9
3	—	—	—	—	3

$$\therefore \quad \text{LCM} = 2 \times 2 \times 2 \times 3 \times 3 = 72$$

The next step is to divide each of the denominators into the LCM (72), these values are then multiplied by the numerators and placed on the top line over the *common* denominator (72), which of course is also the LCM.

$$\frac{45}{72} + \frac{30}{72} + \frac{12}{72} + \frac{18}{72} + \frac{48}{72} = \frac{45 + 30 + 12 + 18 + 48}{72}$$

$$= \frac{153}{72} = 2\tfrac{9}{72} = 2\tfrac{1}{8} \quad Answer$$

It can easily be checked whether or not you have the correct values, by cancelling out each term to its original form. For example $\dfrac{45}{72}$ will cancel out to $\dfrac{5}{8}$ by dividing top and bottom by 9. If any term does not cancel back into its original form there is an error.

Exactly the same procedure is adopted for algebraic terms where an unknown term is represented by a letter.

41)   Example: Find the LCM of a, $8a$ and $9a^2$

a	$a,$	$8a,$	$9a^2$
a	—	8,	$9a$
2	—	8,	9
2	—	4,	9
2	—	2,	9
3	—	—	9
3	—	—	3

$$\text{LCM} = a \times a \times 2 \times 2 \times 2 \times 3 \times 3 = 72a^2 \quad Answer$$

Multiplication of Fractions is a simple matter

$$\frac{3}{7} \times \frac{5}{8} \times \frac{9}{17} = \frac{3 \times 5 \times 9}{7 \times 8 \times 17} = \frac{135}{952}$$

Division of Fractions may be simplified by turning the bottom line upside down and multiplying.

$$\frac{\dfrac{2}{3} \times \dfrac{4}{5}}{\dfrac{7}{8} \times \dfrac{5}{9}} \quad \frac{2}{\underset{1}{3}} \times \frac{4}{5} \times \frac{8}{7} \times \frac{\overset{3}{9}}{5} = \frac{192}{175} = 1\tfrac{17}{175}$$

Ratios

To say that one term is greater than another does not mean very much. For example if 10cm was greater than another unspecified length, the unspecified length could be 2mm or 9cm.

However, if one length is 4cm and the other 20cm, the latter is five times the former while the difference between them is 16cm.

Therefore the ratio of two quantities is the number of times that one quantity is contained in the other. In the example above

$$20:4 \quad \text{or} \quad 5:1$$

This gives us immediate and valuable information.

Proportion

Terms are said to be in proportion when their ratios are equal. e.g.

$$\frac{1}{3} = \frac{5}{15}$$

Therefore 1, 3, 5 and 15 are in proportion. $18:12:10:16$ are *not* in proportion, nor are $9:6:5:8$ but $18:12:10:16$ are proportional to $9:6:5:8$ because $18:9 = 12:6 = 10:5 = 16:8$.

Variation

When two quantities are said to be increasing or decreasing at the same rate; they are said to be directly proportional to, or vary directly as, the other.

If one quantity increases as the other decreases, they are said to be inversely proportional to, or to vary inversely as the other.

$$\therefore a \propto b \quad \text{mean directly as}$$

or

$$a \propto \frac{1}{b} \quad \text{means inversely as}$$

Transposition of Formulae

When a formula is transposed it is put into a different form.

42) Example: $a^2 = b^2 + c^2 - 2bc \cdot \cos A$. (The cosine formula.) If we transpose for a positive $\cos A$, we will have:

$$a^2 + 2bc \cdot \cos A = b^2 + c^2$$

$$2bc \cdot \cos A = b^2 + c^2 - a^2$$

Note: Change the side we change the SIGN. But this does *not* apply when we multiply or divide.

$$\therefore \quad \cos A = \frac{b^2 + c^2 - a^2}{2bc}$$

In other words, we have isolated $\cos A$ from the rest of the formula and put a^2 back into it. In carrying out this sort of operation, all the rules of arithmetic, algebra, etc. *must* be obeyed.

Remember that whatever operation is carried out on one side, *must* be carried out on the other. When several brackets are involved, work from the inside outwards. Get rid of root signs and fractions as soon as possible if you can.

Sometimes there are several different methods of dealing with the same problem, and answers can be written in different ways. Because one answer does not look exactly the same as another for a particular problem, this does not mean that it is wrong.

Simple Equations are equations where the independent variable is raised to the power of one.

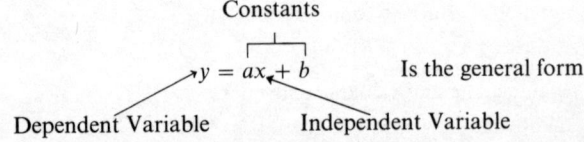

Constants

$$y = ax + b \qquad \text{Is the general form}$$

Dependent Variable   Independent Variable

The dependent variable is so named because it depends on the rest of the equation for its value. If we change the value of the independent variable x we shall automatically get a new value for y.

The power of x tells us how many values we have to find for the answer. In this case the power is one, therefore there will be one value for x in the answer.

If we subtract one from the power, it tells us how many curves there are in the graphical solution.

In this case $1 - 1 = 0$; therefore the graph would be in the form of a straight line.

If the coefficient of $x(a)$ is positive, it indicates that the graph will slope to the right. If it is negative it will slope to the left.

If the constant a is given, it indicates the tangent of the slope of the graph. Remember that the slope of the graph is always measured between the line of the graph and the x axis (the horizontal axis).

The constant b indicates where the line of the graph cuts the y axis (the vertical axis). If it is positive it cuts the vertical above the x axis; if it is negative it cuts below the x axis.

By looking at an equation and remembering the basic principles mentioned above, a student can very easily form a mental picture of how the graphical solution should look before he actually works out anything at all on paper. Such knowledge can prove to be invaluable when time is short, or when a decision has to be made about the scale to use when plotting a graph. For example, is it necessary to plot a graph using all four quadrants or will one do, in which case a larger scale can be used and a more accurate answer obtained. Sometimes a rough sketch in the margin can help and also give a very good approximation of the answer to be expected.

43) Example: Solve $4x + 3 = 39$

$$\text{General form} \quad ax + b = y$$

$$4x + 3 = 39$$

Transpose all the unknowns on one side and all the knowns on the other.

$4x = 39 - 3$ Change the side, change the sign

$4x = 36$

$$x = \frac{36}{4} = 9 \quad \textit{Answer}$$

Using some imagination, or alternatively making a rough sketch it can be seen that: the constant $a = 4 =$ tangent of slope $=$ coefficient of x, is positive and therefore will slope to the right. The constant $b = 3$; therefore the line of the graph will cut the vertical axis above the horizontal axis because it is positive.

There will be a fairly steep slope because $a = 4$; the higher the value of a the steeper the slope.

The shape of the graph will alter according to the scale used. REMEMBER that a protractor can only be used to measure the slope of the graph so long as the scales on both the vertical and the horizontal axes are the same. Otherwise it does not matter what scales are used, it is purely a matter of convenience whether the scale on the y-axis is the same as that used on x-axis.

Simultaneous Equations consist of 2, 3, 4, or more simple equations, with the same number of unknowns as there are equations. There are many ways of dealing with this type of equation and each one must be dealt with according to its merits. The more usual methods are shown below.

44) Solve

$$\text{(a)} \quad 5x + 3y = 41$$

$$\text{(b)} \quad 7x + 6y = 70$$

In using the following method, we first of all have to decide which of the unknowns we are going to evaluate first. The other unknown has to be eliminated.

To eliminate one of the unknowns we have to make their coefficients the same, although their signs may be different. Therefore it may be necessary to multiply either one or both of the equations by some convenient number in order to do this. In this case equation (a) will be multiplied by 2: This will give $10x + 6y = 82$.

Rule: If the signs of the coefficients of the unknown to be eliminated are the SAME— SUBTRACT. If they are DIFFERENT—ADD.

$$10x + 6y = 82$$
$$\underline{7x + 6y = 70}$$

Same—Subtract $\quad 3x \qquad = 12$

$$\therefore \quad x = \frac{12}{3} = 4$$

$x = 4$ is now substituted in either equation (a) or (b); we would naturally choose the easier one.

Substituting $x = 4$ in (a) $5x + 3y = 41$

will give $(5 \times 4) + 3y = 41$

$20 + 3y = 41$

transposing $3y = 41 - 20$

therefore $3y = 21$

dividing by 3 $y = \dfrac{21}{3} = 7$

Answer $x = 4$ when $y = 7$

45)   Alternatively, by substitution. Once again, it does not matter which equation or which unknown is used. From equation (a) $5x + 3y = 41$ by transposition,

$$y = \frac{41 - 5x}{3}$$

Substituting this value of y in equation (b) $7x + 6y = 70$ will give

$$7x + \overset{2}{6} . \left[\frac{41 - 5x}{3} \right] = 70$$

Multiplying through the bracket by 2

$$7x + 82 - 10x = 70$$

Transposing

$$82 - 70 = 10x - 7x$$

$$12 = 3x$$

$$x = \frac{12}{3} = 4$$

By substituting the value of $x = 4$ in (a) as shown above we find that $y = 7$.

Answer $x = 4$ when $y = 7$

46)   Finally, by graphical solution. Solve

(a) $5x + 3y = 41$

(b) $7x + 6y = 70$

By making a rough sketch as mentioned earlier in Simple Equations (page 29) you will see that it is only necessary to use the upper righthand quadrant, also it will be sufficient to use only positive values for x when calculating values of y. Remember that *two* simple equations are to be plotted, and the values of x and y that satisfy both equations at the same time are to be found at the point where the graphs cross. Firstly, transpose for y (with coefficient of 1) with each equation. From (a)

$$5x + 3y = 41; \qquad y = \frac{41 - 5x}{3} = 13\tfrac{2}{3} - 1\tfrac{2}{3}x$$

and from (b)

$$7x + 6y = 70; \qquad y = \frac{70 - 7x}{6} = 11\tfrac{2}{3} - 1\tfrac{1}{6}x$$

By substituting any convenient values for x in these equations corresponding values for y are calculated. Only two points are needed to draw a straight line, but for convenience it is usual to use three as a check. The b value in $y = ax + b$ is one point that could be used, as this value is always on the y-axis when x is zero. To calculate values of y:

(a)

x	0	2	4
$-1\frac{2}{3}x$	0	$-3\frac{1}{3}$	$-6\frac{2}{3}$
$13\frac{2}{3}$	$13\frac{2}{3}$	$13\frac{2}{3}$	$13\frac{2}{3}$
y	$13\frac{2}{3}$	$10\frac{1}{3}$	7

(b)

x	0	2	4
$-1\frac{1}{6}x$	0	$-2\frac{1}{3}$	$-4\frac{2}{3}$
$11\frac{2}{3}$	$11\frac{2}{3}$	$11\frac{2}{3}$	$11\frac{2}{3}$
y	$11\frac{2}{3}$	$9\frac{1}{3}$	7

Each value of x and its corresponding value of y will give a point to be plotted on the graph. Plot the points and draw a straight line for each graph. Where the lines cross, drop a line vertically downwards to get the value for x, and a line horizontally across the page to the y-axis to get the value for y. These two values being the solution to the problem.

It will be noticed in this instance that $x = 4$ and $y = 7$ appears in both tables, hence it is not strictly necessary to plot the graph unless it is specially asked for. However, this is unusual, and in this particular case it is of interest to plot the graph because the two lines are very close together, and it will emphasize the need for accuracy in plotting if a reasonably accurate result is to be obtained (see Fig. 2.2).

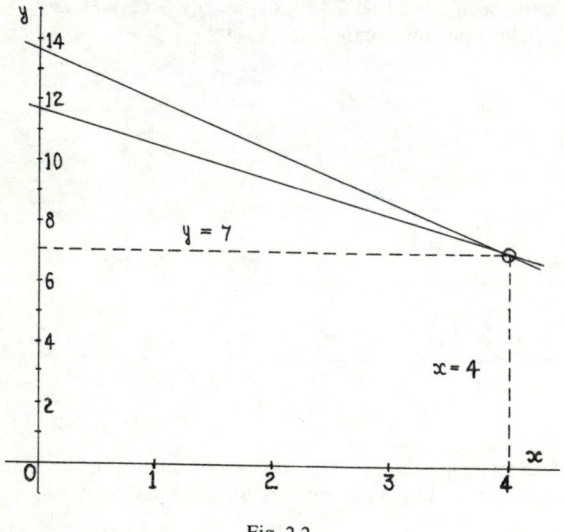

Fig. 2.2

Quadratic Equations of which the general form is

$$y = ax^2 + bx + c$$

The unknown x is to the power of 2; this indicates that there are two values to find for x. Subtracting one from the power of two, $2 - 1 = 1$, indicates that there will be one curve in the graphical solution.

There are four methods for solving quadratic equations, as follows:

(a)   Factorizing
(b)   By completing the square
(c)   Formula
(d)   Graphical solution

As method (a) has already been dealt with in example 39, page 24, it would be a helpful comparison to use the same example for the next three methods.

47)   (b)   Completing the square

$$x^2 + 5x - 36 = 0$$

Of the four methods, this is the only one where it is necessary to reduce the coefficient of x^2 to unity (1). In this instance it is already in the correct form, but it is important to remember that if the coefficient is *more* than *one*, then it has to be reduced to one.

The first thing to do, is to check the equation to see whether or not it is already a Perfect Square, see page 24. In this case it is not, so we can proceed. When we refer to Completing the Square, we are quite literally completing the area of a square, hence its title. The first two terms are always included in the area, but the last term (-36) is not. Therefore the last term has to be transposed so that a more suitable term can be substituted in its place in order to complete the square and therefore make it into a *perfect square*. Now that we have the area of the large square and the area of the two rectangles, it should be clearly seen from Fig. 2.3 that it is only necessary to add the area of the small square to have a perfect square.

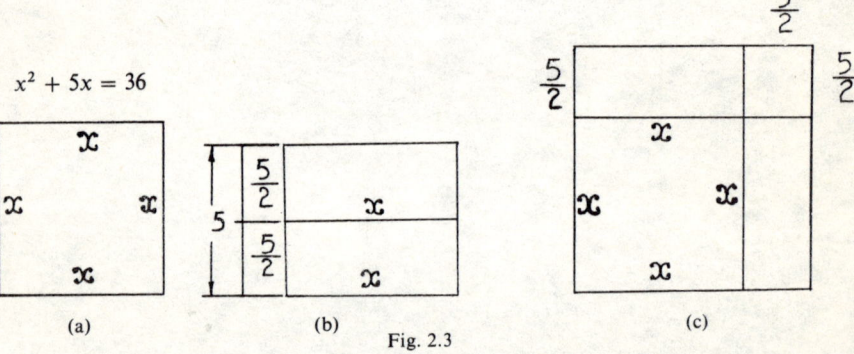

Fig. 2.3

The term $5x$ represents the areas of the two rectangles in the diagram. If $5x$ was represented by one large rectangle we could, in mathematical terms, say that the area was equal to:

$$5 \times x = 5x$$

Unfortunately, this would not fit into the diagram, therefore the large rectangle has been split into two smaller rectangles each representing half the area of the large rectangle. Therefore, in mathematical terms we can write:

$$\left[\frac{5}{2} \times x\right] + \left[\frac{5}{2} \times x\right] = [2\tfrac{1}{2}x + 2\tfrac{1}{2}x] = 5x$$

As the width of the rectangles is equal to the length of the sides of the small square; the area of the small square must be equal to $\left[\dfrac{5}{2} \times \dfrac{5}{2}\right] = \left[\dfrac{5}{2}\right]^2$ or, in words, the area must be equal to [Half the *coefficient* of x (5)]^2.

This area must be added to both sides in order to preserve the balance. The equation will now be:

$$x^2 + 5x + \left[\frac{5}{2}\right]^2 = 36 + \left[\frac{5}{2}\right]^2$$

The lefthand side of the equation is now in the form of a PERFECT SQUARE. As a perfect square, the factors will be the same in both brackets (reference page 25), the equation can now be written in the following form:

$$\left[x + \frac{5}{2}\right]^2 = 36 + \left[\frac{5}{2}\right]^2$$

$$\left[x + \frac{5}{2}\right]^2 = 36 + \frac{25}{4}$$

$$\left[x + \frac{5}{2}\right]^2 = \frac{144 + 25}{4}$$

Taking the square root of both sides

$$x + \frac{5}{2} = \pm\sqrt{\frac{144 + 25}{4}}$$

$$x + \frac{5}{2} = \pm\sqrt{\frac{169}{4}}$$

$$x + \frac{5}{2} = \frac{13}{2}$$

$$\therefore \quad x = -\frac{5}{2} \pm \frac{13}{2}$$

So

$$x = \frac{-5 - 13}{2} = -\frac{18}{2} = -9$$

or

$$x = \frac{-5 + 13}{2} = \frac{8}{2} = 4$$

Answer $x = -9$ or 4

Once again use the same quadratic equation and solve it using the formula method (c). Now try—as an exercise—to complete the square with $ax^2 + bx + c = 0$; you should arrive at

$$x = \frac{-b \pm \sqrt{b^2 - 4ac}}{2a}$$

If you cannot do it, have a look at the solution in Test Paper 4, Question 4;

48) Solve $x^2 + 5x - 36 = 0$ by formula. Call the coefficient of x^2, a; the coefficient of x, b; and the constant, c;

$\left. \begin{array}{l} a = 1 \\ b = 5 \\ c = -36 \end{array} \right\}$ Now substitute these values in the above formula, as follows, and remember: it is the first term that you write where you are most likely to make a mistake.

In the formula there is a $-b$; this tells us that whatever value we have for b, has to be multiplied by -1 before it is substituted in the formula. Therefore $b = +5 \times -1 = -5$.

$$x = \frac{-5 \pm \sqrt{5^2 - (4 \times 1 \times -36)}}{2 \times 1}$$

$$x = \frac{-5 \pm \sqrt{25 + 144}}{2}$$

$$x = \frac{-5 \pm \sqrt{169}}{2}$$

$$x = \frac{-5 \pm 13}{2}$$

$$\therefore \quad x = \frac{-5 - 13}{2} = \frac{-18}{2} = -9$$

or

$$x = \frac{-5 + 13}{2} = \frac{8}{2} = 4$$

Answer $x = -9$ or 4

49) Finally solve the same problem $x^2 + 5x - 36 = 0$ by graphical solution.

As with simple equations, a very good idea of what the graph will look like can be obtained by remembering a few basic rules. In the general formula $ax^2 + bx + c = 0$.

If a is positive the graph will be like a letter u.
If a is negative the graph will be like a letter u upside down.
If c is positive the graph will cut the y-axis *above* the x-axis.
If c is negative the graph will cut the y-axis *below* the x-axis.
If a and b are both positive the graph will move to the left of the y-axis.
If a and b are both negative the graph will move to the left of the y-axis.
If a is positive and b is negative the graph will move to the right of the vertical (y-axis).
If a is negative and b is positive the graph will move to the right of the y-axis.
In other words, if a and b are the same, the graph moves left.
If a and b are different, the graph moves right.

By considering the information given in the equation, we can see that the graph will be the right way up like a letter u. The coefficients of x^2 and x are both positive, therefore we can expect the graph to be more on the lefthand side of the centre. That c is -36 tells us that the graph cuts the y-axis quite a long way below the x-axis. From this information it is obvious that we need to use all four quadrants. In tabulating values for y, we will have to substitute more negative values for x than positive values, because it is more to the left or negative side on the x-axis. Tabulating values for x and y.

x	-10	-8	-6	-4	-2	0	2	4	6
x^2	100	64	36	16	4	0	4	16	36
$-5x$	-50	-40	-30	-20	-10	0	10	20	30
-36	-36	-36	-36	-36	-36	-36	-36	-36	-36
y	14	-12	-30	-40	-42	-36	-22	0	30

Now plot these points and draw in the graph (Fig. 2.4). The required values for x are where the graph cuts the x-axis.

$$Answer \quad x = -9 \text{ or } 4$$

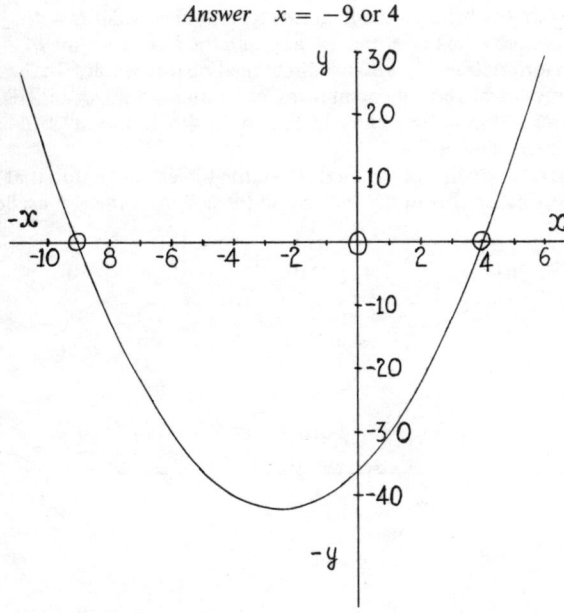

Fig. 2.4

Simultaneous Quadratic Equations consist of a simple or linear equation and a quadratic equation, both of which have already been described. Where the linear graph crosses the quadratic are the two points where the values of x and y satisfy both equations at the same time. Therefore in the answer there will be two values for x, and two values for y.

It is of interest to note that, if the linear equation is added to the quadratic equation, we will get a quadratic equation which can be solved by one of the methods already described.

Cubic Equations are equations where the power of the unknown (x) is 3. This indicates that we have to find *three* values for x; and $3 - 1 = 2$, shows us that there will be two curves in the graphical solution. In the majority of cases, cubic equations have to be solved graphically. If, however, you are asked to solve such an equation mathematically, then it has most likely been specially selected. For these and other more advanced problems, examples have been included in the Test Papers (Part 3).

TRIGONOMETRY means the measurement of triangles and deals with the relationship between the sides and angles of triangles. The trigonometrical ratios, complements and reciprocals, together with various formulae, are to be found in Part 1 of this book.

A great deal of confusion is caused because students do not readily understand exactly what they are dealing with when they have to use sines, tangents and secants.

The word sine means *bay* or *notch*, which in fact is formed within the circle diagram.

Tangent means to *touch*; the tangent touches the circumference of a circle without actually going through it.

Secant means to *cut*; the radius from the centre of the circle actually *cuts* the circumference and continues outwards, therefore it will be appreciated that if the radius of the circle is equal to one, then the value of the secant cannot be less than one.

When taking angles from circle diagrams or graphical solutions, the angles should always be read between the horizontal (x) axis and the radius, or line of the graph.

In the 1st quadrant ($0° - 90°$) take a direct reading from tables. In the 2nd quadrant ($90°-180°$) subtract from $180°$, then enter tables. In the 3rd quadrant ($180°-270°$) from the *angle*, subtract $180°$ and then enter tables. In the 4th quadrant ($270°-360°$) subtract from $360°$ before entering tables.

After taking out the value of the angle from the tables, make sure that the right sign is attached, as in Fig. 2.5. From the triangle so formed from the radius, horizontal axis

Quadrant	Trig. Ratio	Sign
1st	sine	+
	cosine	+
	tangent	+
2nd	sine	+
	cosine	−
	tangent	−
3rd	sine	−
	cosine	−
	tangent	+
4th	sine	−
	cosine	+
	tangent	−

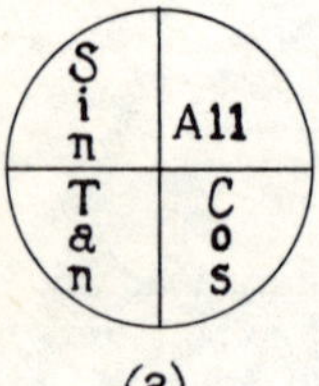

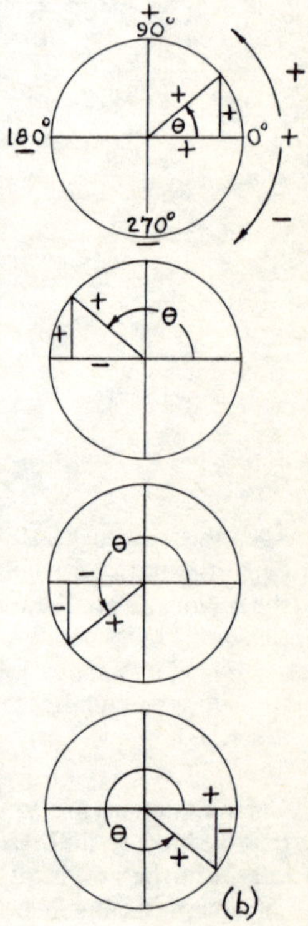

Aid to Memory

(a)

(b)

Fig. 2.5

36

and a vertical line dropped from the radius to the horizontal, attach the following signs:

The radius (hypotenuse of the triangle) is always positive whichever quadrant it is in. If the vertical line is above the horizontal, it is positive; if below, negative.
If the horizontal line is to the right of the vertical, it is positive; if to the left, negative.

Example: In the first quadrant the cosine is positive, as shown below:

$$\cos \theta = \frac{\text{Adjacent}}{\text{Hypotenuse}} = \frac{+}{+} = +$$

In the second quadrant the cosine is negative:

$$\cos \theta = \frac{\text{Adjacent}}{\text{Hypotenuse}} = \frac{-}{+} = -$$

50) A steel support wire 12m long was at an angle of 60° to the ground. How high above the ground was the point where it was fastened to a flagstaff?

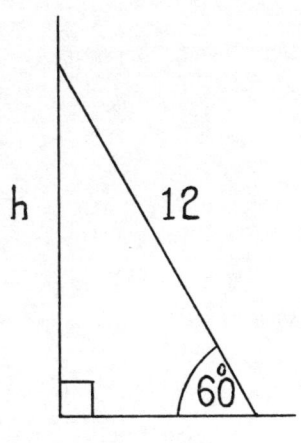

Fig. 2.6

$$\frac{\text{Required side}}{\text{Known side}} = \frac{h}{12} = \text{sine } 60°$$

$$\therefore \quad h = 12 \times \text{sine } 60°$$

$$h = 12 \times 0\!\cdot\!866$$

$$h = 10\!\cdot\!392\text{m} \quad Answer$$

Instead of multiplying by the sine of 60°, we could just as easily have divided by its reciprocal, cosecant 60°.

$$\text{where} \quad h = \frac{12}{\text{cosec } 60°}$$

$$h = \frac{12}{1\!\cdot\!1547}$$

$$\therefore \quad h = 10\!\cdot\!392\text{m} \quad Answer$$

The Sine Rule

$$\frac{a}{\sin A} = \frac{b}{\sin B} = \frac{c}{\sin C}$$

In this rule it is necessary to have any three parts of a triangle to find a fourth part; this means that we use any two pairs of the three shown in the above formula, provided that the sides and angles used are opposite one another. This rule is easy to use and has the advantage of the sine being positive in both first and second quadrants, therefore we do not have to worry about negative angles.

51) Solve the triangle

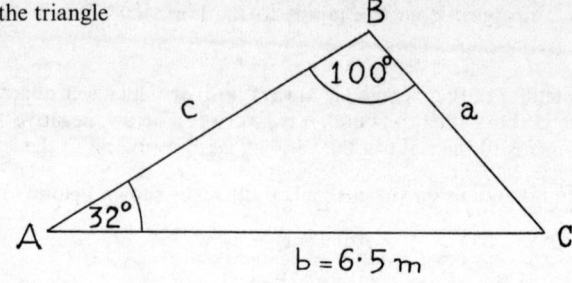

Fig. 2.7

$$\frac{a}{\sin A} = \frac{b}{\sin B}$$

To find side a,

$$a \cdot \sin B = b \cdot \sin A \quad \therefore \quad a = \frac{b \cdot \sin A}{\sin B}$$

$$a = \frac{6 \cdot 5 \times \sin 32°}{\sin 100°} \quad \text{(Remember } \sin 100° = 180° - 100° = \sin 80°)$$

$$a = \frac{6 \cdot 5 \times 0 \cdot 5299}{0 \cdot 9848} = \frac{3 \cdot 44435}{0 \cdot 9848}$$

$$a = 3 \cdot 497 \text{m} \quad Answer$$

To find side c.

$$\text{Angle } C = 180° - (\text{Angle } A + \text{Angle B})$$

$$\angle C = 180° - (32° + 100°)$$

$$\angle C = 180° - 132°$$

$$\angle C = 48°$$

$$\frac{c}{\sin C} = \frac{a}{\sin A} \quad \therefore \quad c = \frac{a \cdot \sin C}{\sin A}$$

$$c = \frac{3 \cdot 497 \times \sin 48°}{\sin 32°}$$

$$c = \frac{3 \cdot 497 \times 0 \cdot 7431}{0 \cdot 5299}$$

$$c = \frac{2 \cdot 5986}{0 \cdot 5299} = 4 \cdot 904 \text{m}$$

$$c = 4 \cdot 904 \text{m} \quad Answer$$

The Cosine Rule

$$a^2 = b^2 + c^2 - 2bc \cdot \cos A$$
$$b^2 = a^2 + c^2 - 2ac \cdot \cos B$$
$$c^2 = a^2 + b^2 - 2ab \cdot \cos C$$

To transpose for cos A, see example 42.

The cosine formula is perhaps a little more difficult to deal with mathematically because, if an angle greater than 90° is used, we go into the second quadrant and it is then necessary to use a negative angle. This then changes the minus sign in the cosine rule to a positive one. See below:

$$a^2 = b^2 + c^2 - 2bc \cdot \cos A \qquad \text{(Positive cos } A)$$

Now suppose we have a negative cos A, we will get: positive

$$a^2 = b^2 + c^2 - 2bc \times -\cos A \qquad \text{(Where } -A = 180° - A)$$

$$- \times - = + \quad \text{and} \quad \rightarrow \quad \text{(Multiplication always takes priority)}$$

$$\therefore \quad a^2 = b^2 + c^2 + 2bc \cdot \cos(180° - A)$$

52) Using the same triangle as for example 51 (Sine Rule), we will use different sides. As will be seen, when using the cosine rule the *included* angle must be used. An included angle is an angle which is included *between* two sides of a triangle, as opposed to being opposite.

To find side a; Given $\angle A = 32°$, side $b = 6.5$m and side $c = 4.904$m

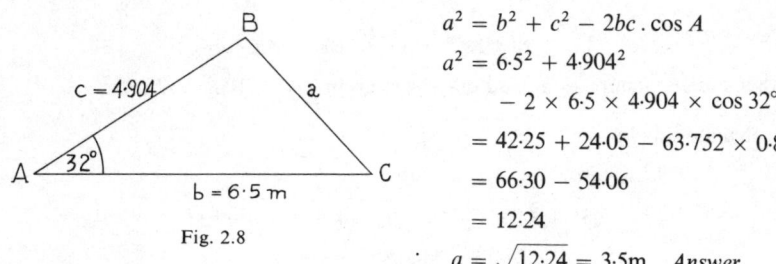

Fig. 2.8

$$a^2 = b^2 + c^2 - 2bc \cdot \cos A$$
$$a^2 = 6.5^2 + 4.904^2$$
$$\qquad - 2 \times 6.5 \times 4.904 \times \cos 32°$$
$$= 42.25 + 24.05 - 63.752 \times 0.8480$$
$$= 66.30 - 54.06$$
$$= 12.24$$
$$\therefore \quad a = \sqrt{12.24} = 3.5\text{m} \quad \textit{Answer}$$

To find $\angle C$;

$$\cos C = \frac{a^2 + b^2 - c^2}{2ab}$$

$$= \frac{3.5^2 + 6.5^2 - 4.904^2}{2 \times 3.5 \times 6.5}$$

$$= \frac{12.25 + 42.25 - 24.05}{45.5}$$

$$\therefore \quad \cos C = \frac{30.45}{45.5} = 0.6692$$

$$\angle C = 48° \quad \textit{Answer}$$

It is not necessary to calculate $\angle B$, because we only have to subtract $\angle A + \angle C$ from $180°$.

$$\therefore \quad \angle B = 180° - (32° + 48°)$$
$$= 180° - 80°$$
$$\angle B = 100°$$

However, because $\angle B$ exceeds $90°$, it will be a good exercise to calculate it. As will be seen, a negative angle will be obtained. Therefore to obtain a positive angle we shall subtract it from $180°$, which as can be seen from above, should be $100°$.

$$\cos B = \frac{a^2 + c^2 - b^2}{2ac}$$
$$= \frac{3·5^2 + 4·904^2 - 6·5^2}{2 \times 3·5 \times 4·904}$$
$$\cos B = \frac{12·25 + 24·05 - 42·25}{2 \times 3·5 \times 4·904}$$
$$= \frac{36·30 - 42·25}{34·328}$$
$$= \frac{-5·95}{34·328} = -0·1733$$
$$\cos B = -0·1733$$
$$\angle B = -80°$$
$$\therefore \quad \angle B = 180° - 80° = 100° \quad Answer$$

As a further exercise, suppose we had only been given $\angle B = 100°$, and sides $a = 3·5$, $c = 4·904$m.

To calculate side b;

$$b^2 = a^2 + c^2 - 2ac \cdot \cos B$$
$$b^2 = 3·5^2 + 4·904^2 - 2 \times 3·5 \times 4·904 \times \cos 100°$$
$$= 12·25 + 24·05 - 34·328 \times -\cos 80°$$
$$= 36·30 - 34·328 \times -0·1736$$
$$= 36·30 + 9·959$$
$$= 42·259$$
$$\therefore \quad b = \sqrt{42·259} = 6·5\text{m} \quad Answer$$

Identities

By using Pythagoras, and dividing $a^2 = b^2 + c^2$ by each term in turn, i.e. $(a^2, b^2$ and $c^2)$ we can develop nine standard identities.

$$\frac{a^2}{a^2} = \frac{b^2}{a^2} + \frac{c^2}{a^2}$$

By substituting trig. ratios we get

$$1 = \sin^2 \theta + \cos^2 \theta$$

and by transposition,

$$1 - \sin^2 \theta = \cos^2 \theta$$

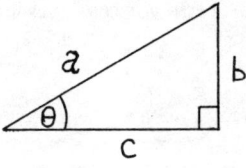

Fig. 2.9

also

$$1 - \cos^2 \theta = \sin^2 \theta$$

$$\frac{a^2}{b^2} = \frac{b^2}{b^2} + \frac{c^2}{b^2}$$ substituting trig. ratios

we get

$$\text{cosecant}^2 \theta = 1 + \text{cotangent}^2 \theta$$

by transposition

$$\text{cosecant}^2 \theta - 1 = \text{cotangent}^2 \theta$$

also

$$\text{cosecant}^2 \theta - \text{cotangent}^2 \theta = 1$$

$$\frac{a^2}{c^2} = \frac{b^2}{c^2} + \frac{c^2}{c^2}$$ substituting trig. ratios

we get

$$\text{secant}^2 \theta = \text{tangent}^2 \theta + 1$$

by transposition

$$\text{secant}^2 \theta - \text{tangent}^2 \theta = 1$$

also

$$\text{secant}^2 \theta - 1 = \text{tangent}^2 \theta$$

GEOMETRY: SOLIDS, SURFACES, PLANES AND LINES

Solids   are bounded by surfaces and have definite shapes.
Surfaces   are either plane (flat) or curved, they have no thickness.
Lines   are determined by the intersection of two surfaces. Lines have no thickness nor breadth.
Points   are determined by the intersection of two lines.

When two lines intersect they form *angles*, as follows:

41

When two adjacent lines intersect any two adjacent angles add up to 180°

$$\propto + \theta = 180°$$

and opposite angles are equal

$$\angle \propto = \angle \propto$$
$$\angle \theta = \angle \theta$$

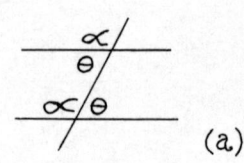

Fig. 2.10

Parallel straight lines are defined as lines which never meet, no matter how far they are produced. A line which cuts two parallel lines is known as a transversal, which gives rise to the following conditions:

(a) corresponding angles are equal

$$\angle \propto = \angle \propto$$

(b) alternate angles are equal

$$\angle \theta = \angle \theta$$

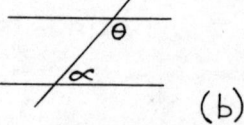

(a)

(c) The two angles contained by two parallel lines and the transversal add up to 180°

$$\angle \theta + \angle \propto = 180°$$

(b)

Fig. 2.11

When two straight lines cut by another, in such a manner that one of the above conditions exists, then the two straight lines are said to be parallel. When two straight lines form the following angles, they are said to be

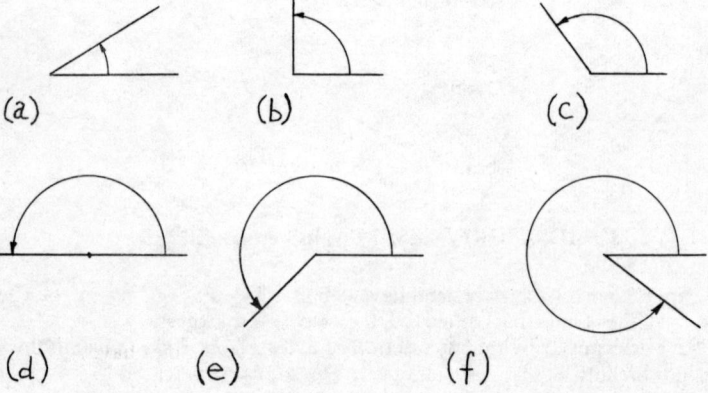

Fig. 2.12 (a) Acute Angle (b) Right Angle (c) Obtuse Angle (d) Straight Line (e) Reflex Angle (f) Reflex Angle

Triangles

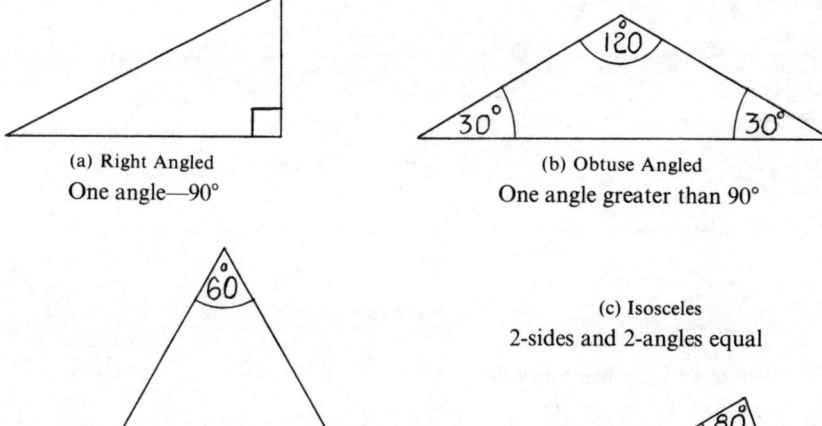

(a) Right Angled
One angle—90°

(b) Obtuse Angled
One angle greater than 90°

(c) Isosceles
2-sides and 2-angles equal

(d) Equilateral
All sides and angles equal

(e) Acute Angled
All angles less than 90°

Fig. 2.13

Congruent Triangles are alike in shape and size

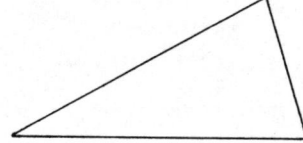 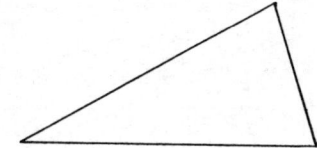

Fig. 2.14

Triangles can only be said to be congruent if the following conditions can be proved to exist:

(a) Two sides and the included angle are equal.
(b) Three sides are equal.
(c) Two angles and a corresponding side are equal.

Similar Triangles are alike in shape but *not* in size.

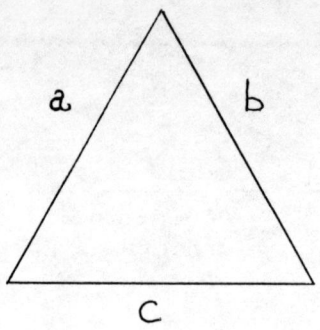
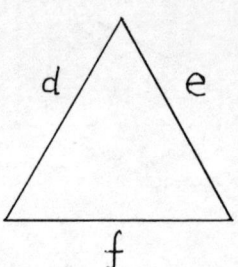

Fig. 2.15

The ratios of the sides being as follows:

$$\frac{c}{f} = \frac{a}{d} = \frac{b}{e}$$

Note that the corresponding sides have equal opposite angles.

53)   Construct a triangle with a base diameter of 8cm and a slant height of 12cm. Calculate the height of its centroid above the base, and then erect a perpendicular to one of the slant sides in such a manner that it passes through the centroid. Find by similar triangles:

(a)   The length of the line drawn between the slant side and the centroid.
(b)   The distance down the slant side from the top of the cone to the point where the perpendicular was erected.

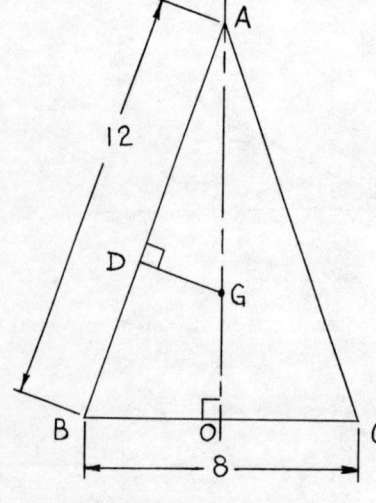

Fig. 2.16

$$AO = \sqrt{12^2 - 4^2}$$
$$= \sqrt{144 - 16}$$
$$= \sqrt{128} = 11\cdot31cm$$
$$OG = \frac{11\cdot31}{3} = 3\cdot77cm$$

(Divide by 3 for an area or 4 for a solid)

$$\therefore \quad AG = 11\cdot31 - 3\cdot77 = 7\cdot54cm$$

By similar triangles

$$\frac{GD}{BO} : \frac{AG}{AB} \quad \therefore \quad \frac{GD}{4} : \frac{7\cdot54}{12}$$

$$GD = \frac{7\cdot54 \times 4}{12} = 2\cdot51cm \quad Answer\ (a)$$

and as

$$\frac{AD}{AO} : \frac{GD}{BO} \qquad \therefore \quad \frac{AD}{11 \cdot 31} : \frac{2 \cdot 51}{4}$$

$$\therefore \quad AD = \frac{2 \cdot 51 \times 11 \cdot 31}{4} = \frac{28 \cdot 388}{4} = 7 \cdot 097 \text{cm} \quad \textit{Answer} \text{ (b)}$$

$$\textit{Answer} \quad GD = 2 \cdot 51 \text{cm}$$

$$AD = 7 \cdot 097 \text{cm}$$

PART 3

Test Papers and Solutions

In this section there are twelve test papers, each incorporating nine questions. The solutions follow immediately after each test paper.

The questions have been arranged so as to provide the student with a test paper, as shown, with which he can test his mathematical ability by attempting six questions over a period of three hours.

Alternatively, should the student wish to attempt a series of log questions he can attempt Question 1 from each paper or, for trigonometry, Question 6.

Question 7 from each paper will provide a series of different types of questions without having to attempt a complete test paper.

The order in which the questions are given is shown below. Remember that there is often more than one solution to a question. The solutions given have been chosen—in some cases—to give the student practice in mathematics, rather than the shortest method.

Question 1 Indices
Question 2 Logarithms
Question 3 Transposition
Question 4 Algebra
Question 5 Graphs (including Simpson's Rules)
Question 6 Trigonometry
Question 7 Various
Question 8 Mensuration
Question 9 Geometry

TEST PAPER 1

Q.1 Evaluate by the use of indices and *without* using tables

(a) $(2^{\sqrt{2}})^{\sqrt{2}}$ (b) $\dfrac{3^{\frac{3}{4}} \times 2^5}{9^{\frac{5}{4}} \times 8}$ (c) $\sqrt[3]{4^2} \times 4^{\frac{3}{4}} \times (\frac{1}{4})^{-\frac{3}{4}}$

Q.2 Given the equation

$$\phi = 1\cdot0565 \log_e \frac{t}{273} + 9 \times 10^{-7} \times \left[\frac{t^2}{2} - 503t\right] + 0\cdot0902$$

Calculate without using a slide rule the value of ϕ if $t = \theta + 273$ when $\theta = 53°$.

Q.3 Transpose for K in

$$x = \frac{c}{k}(\sqrt{1 + k^2t^2} - 1)$$

Q.4 Solve for a and b in

(i) $a^2 + b^2 = 185$

(ii) $a + b = 17$

Q.5 Given $y = x^3 + 2x^2 - 11x + 36$ and using any rule except the mid-ordinate rule, find the swept volume of the area rotated about the x-axis.

Q.6 326 drops of water—spherical in shape—and 0·125 units in diameter, were dropped into a conical shaped wine glass whose diameter is equal to the height of the liquid. Find the height of the liquid.

Q.7 Find the value of θ between 0° and 360° for the following equations.

(a) $\cos 3\theta - \cos \theta = 0$

(b) $3 \sin^2 \theta - 2 \cos \theta = 2$

Q.8 A ship's rectangular tank of 15m length, 9m breadth and 5m depth, has a semi-cylindrical bottom. The total depth of the tank is 9·5m. Find the surface area and volume of the tank.

Q.9 Prove that the radius R of an inscribed circle in any triangle is given by

$$R = \frac{2 \times \text{Area of Triangle}}{\text{Perimeter of Triangle}}$$

Hence calculate the radius of an inscribed circle in a triangle of sides 2, 3 and 4cm respectively.

A.1 (a) $(2^{\sqrt{2}})^{\sqrt{2}} = 2^2 = 4$ *Answer* (a)

(b) $\dfrac{3^{\frac{3}{4}} \times 2^5}{9^{\frac{5}{4}} \times 8} = \dfrac{3^{\frac{3}{4}} \times 2^5}{9^{\frac{5}{4}} \times 2^3}$

$$= \frac{2^{5-3}}{9^{\frac{5-2}{6}}} = \frac{2^2}{9^{\frac{1}{2}}} = \frac{4}{3} = 1\frac{1}{3} \text{ \textit{Answer} (b)}$$

(c)
$$\sqrt[3]{4^2} \times 4^{\frac{3}{4}} \times (\tfrac{1}{4})^{-\frac{3}{4}} = 4^{\frac{2}{3}} \times 4^{\frac{3}{4}} \times \frac{1}{(\frac{1}{4})^{\frac{3}{4}}}$$

$$= 4^{\frac{2}{3}} \times 4^{\frac{3}{4}} \times 4^{\frac{3}{4}}$$

$$= 4^{\frac{8}{4}} = 4^2 = 16 \quad \text{Answer } (c)$$

A.2 $\phi = 1 \cdot 0565 \log_e \dfrac{t}{273} + 9 \times 10^{-7}\left(\dfrac{t^2}{2} - 503t\right) + 0 \cdot 0902$

$\therefore \quad t = \theta + 273 = 53 + 273 = 326 \quad$ Substituting

$$\overset{①}{} \qquad \overset{②}{} \qquad \overset{③}{}$$

$$\phi = 1 \cdot 0565 \log_e \frac{326}{273} + 9 \times 10^{-7}\left(\frac{326^2}{2} - 503 \times 326\right) + 0 \cdot 0902$$

①	②	③
$\log 326 = 2 \cdot 5132$	$\log 326 = 2 \cdot 5132$	$\log 503 = 2 \cdot 7016$
$\log 273 = 2 \cdot 4362$	$ 2$	$\log 326 = 2 \cdot 5132$
$\log t = \overline{0 \cdot 0770}$	$\overline{5 \cdot 0264}$	$\overline{5 \cdot 2148}$
Antilogging $= \underline{1 \cdot 194}$	$\log 2 = 0 \cdot 3010$	Antilogging $= \underline{164,000}$
	$\log \dfrac{326^2}{2} = 4 \cdot 7254$	
	Antilogging $= \underline{53,140}$	

$$\phi = 1 \cdot 0565 \times \underbrace{\log_e 1 \cdot 194}_{④} + \underbrace{9 \times 10^{-7}(53,140 - 164,000)}_{⑤} + 0 \cdot 0902$$

$$\phi = 1 \cdot 0565 \times \underbrace{0 \cdot 1775}_{④} - \underbrace{9 \times 10^{-7} \times 110,860}_{⑤} + 0 \cdot 0902$$

④	⑤
$\log 1 \cdot 0565 = 0 \cdot 0239$	$\log 9 = 0 \cdot 9542$
$\log 0 \cdot 1775 = \overline{1} \cdot 2492$	$\log 10^{-7} = \overline{7} \cdot 0000$
$\overline{1} \cdot 2731$	$\log 110,860 = 5 \cdot 0446$
Antilogging $= \underline{0 \cdot 1875}$	$\overline{2} \cdot 9988$
	Antilogging $= \underline{0 \cdot 09972}$

$$\phi = 0 \cdot 1875 - 0 \cdot 0997 + 0 \cdot 0902$$

$$\phi = 0 \cdot 2777 - 0 \cdot 0997$$

$\therefore \quad \phi = 0 \cdot 1780 \quad \text{Answer}$

A.3

$$x = \frac{c}{k}\left(\sqrt{1 + k^2 t^2} - 1\right)$$

$$x \cdot \frac{k}{c} = \left(\sqrt{1 + k^2 t^2} - 1\right)$$

$$x \cdot \frac{k}{c} + 1 = \sqrt{1 + k^2 t^2}$$

$$\left(x \cdot \frac{k}{c} + 1\right)^2 = 1 + k^2 t^2$$

$$x^2 \cdot \frac{k^2}{c^2} + 2 \cdot x \cdot \frac{k}{c} + \cancel{1} = \cancel{1} + k^2 t^2$$

$$x^2 \cdot \frac{k^2}{c^2} + 2 \cdot x \cdot \frac{k}{c} = k^2 t^2$$

$$2x \cdot \frac{k}{c} = k^2 t^2 - k^2 \cdot \frac{x^2}{c^2}$$

$$2x \frac{k}{c} = k^2 \left(t^2 - \frac{x^2}{c^2}\right)$$

$$2\frac{x}{c} = \frac{k^2}{k}\left(t^2 - \frac{x^2}{c^2}\right)$$

$$2\frac{x}{c} = k\left(t^2 - \frac{x^2}{c^2}\right)$$

$$k = \frac{2\dfrac{x}{c}}{(t^2 - x^2/c^2)}$$

$$k = 2\frac{x}{c}\left(\frac{1}{t^2} - \frac{c^2}{x^2}\right) \quad \textit{Answer}$$

A.4 (i) $a^2 + b^2 = 185$

(ii) $a + b = 17$

From (ii) $a = 17 - b$; Substitute in (i)

$$(17 - b)^2 + b^2 = 185$$

$$289 - 34b + b^2 + b^2 = 185$$

$$2b^2 - 34b + 289 - 185 = 0$$

$$2b^2 - 34b + 104 = 0$$

$$b^2 - 17b + 52 = 0$$

Factorizing

$$b^2 - 4b - 13b + 52 = 0 \qquad \left\{ \begin{array}{l} P = 52b^2 \\ S = -17b \end{array} \right\}$$

$$b(b - 4) - 13(b - 4) = 0$$

$$(b - 13)(b - 4) = 0$$

$$\left. \begin{array}{l} \text{If} \quad b - 13 = 0; \quad \text{then } b = 13 \\[4pt] \text{if} \quad b - 4 = 0; \quad \text{then } b = 4 \end{array} \right\}$$

OR

By substituting $b = 4$ in (ii)

$$a + 4 = 17$$

$$\therefore \quad a = 17 - 4 = 13$$

and

$$a + 13 = 17$$

$$a = 17 - 13 = 4$$

$$\therefore \quad a = 4; \quad b = 13$$
$$a = 13; \quad b = 4 \quad \Big\} \quad Answer$$

A.5 $y = x^3 + 2x^2 - 11x + 36$

x	x^2	x^3	$x^3 + 2x^2 - 11x + 36$	y
−2	4	−8	−8 + 8 + 22 + 36	58
−1	1	−1	−1 + 2 + 11 + 36	48
0	0	0	0 0 0 + 36	36
1	1	1	1 + 2 − 11 + 36	28
2	4	8	8 + 8 − 22 + 36	30

Using Simpson's 1st Rule $h/3(1 + 4 + 2 + 4 + 1)$ where $h = 1$

Ordinate (Radii)	Area πr^2	S.M's	Products
58	π . 3364	1	π . 3364
48	π . 2304	4	π . 9216
36	π . 1296	2	π · 2592
28	π . 784	4	π . 3136
30	π . 900	1	π . 900
			π . 19,208

$$\frac{h}{3} \times \pi \times 19{,}208 = \frac{60\,344}{3} = 20\,114 \cdot 6\overline{6}$$

$$\therefore \quad \text{Volume} = 20\,115 \text{ Units}^3 \quad Answer$$

As a useful exercise, plot the graph using tabulated values for x and y.

A.6 Volume of sphere $= \frac{4}{3}\pi r^3$ or $\frac{\pi}{6}D^3$

Volume of cone $= \frac{1}{3}\pi r^2 h$ or $\frac{\pi D^2}{12} \cdot h$

$$\therefore \quad 326 \times \frac{\pi}{6}D^3 = \frac{\pi D^2}{12} \times h$$

Substitute D for h

$$326 \times \frac{\pi}{6} \times 0 \cdot 125^3 = \frac{\pi D^3}{12}$$

$$\therefore \quad \pi D^3 = 12 \times 326 \times \frac{\pi}{6} \times 0 \cdot 125^3$$

$$D^3 = \overset{2}{\cancel{1}2} \times 326 \times \frac{\pi}{6} \times 0 \cdot 125^3 \times \frac{1}{\pi}$$

$$D^3 = 652 \times 0\cdot125^3$$
$$D = \sqrt[3]{652 \times 0\cdot125^3}$$
$$D = \sqrt[3]{652} \times 0\cdot125$$

Answer $D = 1\cdot084$ Units

$$\log 652^{\frac{1}{3}} = \frac{2\cdot8142}{3}$$
$$= 0\cdot9382$$
$$\log 0\cdot125 = \overline{1}\cdot0969$$
$$\overline{0\cdot0350}$$
Antilogging $= \underline{1\cdot084 \text{ units}}$

A.7 (a)

$$\cos 3\theta - \cos \theta = 0$$
$$\cos 2\theta = 0$$
$$2\cos^2 \theta - 1 = 0$$
$$\cos^2 \theta = \tfrac{1}{2}$$

$$\begin{array}{c|c} - & + \\ \hline - & + \end{array} \qquad \cos \theta = \pm\sqrt{\tfrac{1}{2}} = \pm 0\cdot7071$$

$$\therefore \quad \theta = 45°, 135°, 225° \text{ and } 315° \quad \textit{Answer}$$

(b)
$$3\sin^2 \theta - 2\cos \theta = 2$$
$$3(1 - \cos^2 \theta) - 2\cos \theta = 2$$
$$3 - 3\cos^2 \theta - 2\cos \theta = 2$$
$$3\cos^2 \theta + 2\cos \theta - 1 = 0$$

$$\left.\begin{array}{l} a = 3 \\ b = 2 \\ c = -1 \end{array}\right\} \quad \cos \theta = \frac{-2 \pm \sqrt{2^2 - (4 \times 3 \times -1)}}{2 \times 3}$$

$$\cos \theta = \frac{-2 \pm \sqrt{4 + 12}}{6}$$

$$\cos \theta = \frac{-2 \pm \sqrt{16}}{6} = \frac{-2 \pm 4}{6}$$

$$\therefore \quad \theta = \frac{-6}{6} = -1 = 180°$$

OR

$$\theta = \frac{2}{6} = 0\cdot3\overline{3} = 70°\,32'$$

$$\theta = 70°\,32', 180°, 289°\,28' \quad \textit{Answer}$$

51

A.8   Area of rectangular section $= 2[(9 \times 5) + (15 \times 5)] + (9 \times 15)$

$$= 2[45 + 75] + 135$$
$$= 240 + 135 = 375m^2$$

Area of semi-circular ends $= \pi r^2 = \pi \times 4.5^2 = 63.64m^2$

Area of semi-cylindrical bottom $= \pi r l = \pi \times 4.5 \times 15 = 212.25m^2$

$$Total\ Surface\ Area = 375 + 63.64 + 212.25$$
$$= 650.89m^2$$

$$Volume = (15 \times 9 \times 5) + \left(\frac{\pi r^2}{2} \times l\right)$$

$$= 675 + \left(\frac{\pi \times 4.5^2}{2} \times 15\right)$$

$$= 675 + 477.3$$

$$= 1152.3m^3$$

$$\left. \begin{array}{l} Total\ Surface\ Area = 650.89m^2 \\ Answer\quad Total\ Volume\quad = 1152.3m^3 \end{array} \right\}$$

A.9 (a)

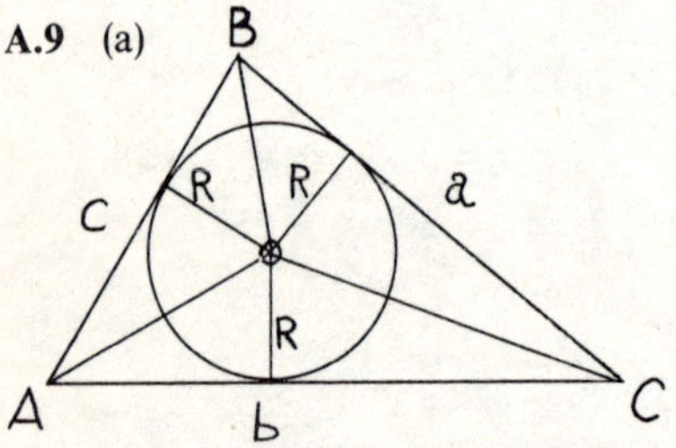

Fig. 3.1

To prove

$$R = \frac{2\ area\ \triangle ABC}{Perimeter}$$

$$= \frac{2\ area\ \triangle ABC}{a + b + c}$$

$$Area\ of\ \triangle BOC = \frac{R}{2}.a$$

$$Area\ of\ \triangle AOC = \frac{R}{2}.b$$

$$Area\ of\ \triangle BOA = \frac{R}{2}.c$$

$$Total\ Area\ \triangle ABC = \frac{R}{2}.(a + b + c)$$

$$\therefore\ \ R = \frac{2.\ area\ \triangle ABC}{a + b + c}\quad Answer\ (a)$$

(b)   Area of $\triangle ABC = \sqrt{s(s - a)(s - b)(s - c)}$

where

$$s = \frac{a + b + c}{2} = \frac{3 + 4 + 2}{2} = 4.5cm$$

$$\text{Area } \triangle ABC = \sqrt{4.5 \times (4.5 - 3)(4.5 - 4)(4.5 - 2)}$$
$$= \sqrt{4.5 \times 1.5 \times 0.5 \times 2.5}$$
$$= \sqrt{8.437} = 2.905 cm^2$$

$$\therefore \quad R = \frac{2.905}{s} = \frac{2.905}{4.5} = 0.6455 \text{cm} \quad \textit{Answer } (b)$$

TEST PAPER 2

Q.1 Simplify

(a) $\quad \sqrt[5]{a^{-3}\sqrt{b^5}} \div \sqrt[4]{b\sqrt[5]{a}}$

(b) $\quad \dfrac{1}{\sqrt[3]{(\sqrt[4]{a^{-\frac{2}{3}}b^{\frac{1}{2}}})^2}}$

(c) $\quad \dfrac{\sqrt{2^8} \times 27^{-3} \times 81 \times 3^8}{3^2}$

Q.2 Evaluate using common logarithms

(a) $\quad \dfrac{1.859 \times 26.27^3 \times 10.69^{\frac{3}{4}} \times 105.6^{-1}}{89.68}$

(b) Evaluate using Napierian logarithms
$$895.2 \times 76.86^{\frac{2}{3}} \times 0.004234^{\frac{2}{3}}$$

Q.3 Transpose for r in
$$Z = \frac{rwl}{\sqrt{r^2 + w^2l^2}}$$

Q.4 Solve

$$\text{(i)} \quad \frac{1}{x} + \frac{1}{y} = 1 \qquad \text{(ii)} \quad 2x + 2y = 9$$

Q.5 A cone has a height h and a base radius r; the slant height is at an angle $\theta°$. Make a formula to find the volume in terms of r and θ and then plot a graph between $0°$ and $80°$.

Q.6 A man standing 4m above sea level sights a mast-top at an angle of $30°28'$. He is 50m from the mast. If the man moves away in the same horizontal plane until the angle is $20°58'$, find the distance the man moved and the height of the mast above sea-level.

Q.7 An orange including peel, may be considered as a perfect sphere, with radius r; the peel being of uniform thickness t. Show that the fraction of original volume removed when the orange is peeled is given by $1 - \left(1 - \dfrac{t}{r}\right)^3$. If the peel is thin so that $\dfrac{t}{r}$ is a small fraction, show that the above expression is equal to $3\dfrac{t}{r}$ approximately. Given $r = 50 \times t$ find the percentage error which would occur using the approximate expression.

Q.8 Two spheres 8cm and 4cm diameter respectively, are placed in a container 10cm diameter. Find the volume of water to just cover them. If the 4cm diameter sphere is removed, find the new height of water.

Q.9 A straight line OAB cuts a circle at A, another line OCD cuts the circle at C. If $BD = OB$ prove that $AC = OC$.

A.1 (a)
$$\sqrt[5]{a^{-3}\sqrt{b^5}} \div \sqrt[4]{b\sqrt[5]{a}} = \frac{[a^{-3}(b^5)^{\frac{1}{2}}]^{\frac{1}{5}}}{(b.a^{\frac{1}{5}})^{\frac{1}{4}}}$$

$$= \frac{(a^{-3}b^{\frac{5}{2}})^{\frac{1}{5}}}{(b^{\frac{1}{4}}a^{\frac{1}{20}})}$$

$$= \frac{(a^{-\frac{3}{5}}b^{\frac{1}{2}})}{(b^{\frac{1}{4}}a^{\frac{1}{20}})}$$

$$= \frac{b^{\frac{1}{2}-\frac{1}{4}}}{a^{\frac{3}{5}+\frac{1}{20}}}$$

$$= \frac{b^{\frac{1}{4}}}{a^{\frac{6+1}{10}}} = \frac{b^{\frac{1}{4}}}{a^{\frac{7}{10}}} \quad \textit{Answer (a)}$$

(b)
$$\frac{1}{\sqrt[3]{(\sqrt[4]{a^{-\frac{2}{3}}b^{\frac{1}{2}}})^2}} = \frac{1}{(\sqrt[4]{a^{-\frac{2}{3}}b^{\frac{1}{2}}})^{\frac{2}{3}}}$$

$$= \frac{1}{(a^{-\frac{2}{3}}b^{\frac{1}{2}})^{\frac{2}{12}}}$$

$$= \frac{1}{a^{-\frac{1}{9}}b^{\frac{1}{12}}} = \frac{a^{\frac{1}{9}}}{b^{\frac{1}{12}}} \quad \textit{Answer (b)}$$

(c)
$$\frac{\sqrt{2^8} \times 27^{-3} \times 81 \times 3^8}{3^2} = \frac{2^{\frac{8}{2}} \times 81 \times 3^8}{3^2 \times 27^3}$$

$$= \frac{2^4 \times 81 \times 3^6}{27^3}$$

$$= \frac{2^4 \times 3 \times \cancel{27} \times \cancel{3^6}}{\cancel{27} \times \cancel{3^2} \times \cancel{3} \times \cancel{3^2} \times \cancel{3}}$$

$$= \frac{16 \times 3}{1}$$

$$= 48 \quad \textit{Answer}$$

A.2 (a)
$$\frac{1.859 \times 26.27^3 \times 10.69^{\frac{1}{3}} \times 105.6^{-1}}{89.68} = \frac{1.859 \times 26.27^3 \times 10.69^{\frac{1}{3}}}{89.68 \times 105.6}$$

No.		
$1.859 = 0.2693$		$\log 26.27 = 1.4194$
$26.27^3 = 4.2582$		$\underline{\hspace{2cm}} 3$
$10.69^{\frac{1}{3}} = 0.3430$		$\log 26.27^3 = \overline{4.2582}$
$\log$ top line $\underline{4.8705}$		
$\log$ b't'mline $= 3.9764$		$\log 10.69 = 1.0290$
$\log$ answer $\overline{0.8941}$		$\underline{\hspace{2cm}} 3$
		$\log 10.69^{\frac{1}{3}} = 0.3430$
		$\log 89.68 = 1.9527$
		$\log 105.6 = 2.0237$
		$\log$ b't'mline $\overline{3.9764}$

Antilogging $= 7.836$ *Answer (a)*

(b) $895 \cdot 2 \times 76 \cdot 86^{\frac{2}{8}} \times 0 \cdot 004234^{\frac{2}{3}}$

$\log_e 895 \cdot 2 = 6 \cdot 7971$
$\log_e 76 \cdot 86^{\frac{2}{8}} = 3 \cdot 7992$
$\log_e 0 \cdot 004234^{\frac{2}{3}} = \overline{4} \cdot 3568$
$\overline{6 \cdot 9531}$

Antilogging $\quad 6 \cdot 9078$

log answer $\quad \overline{0 \cdot 0453}$

$= 1046$

$\log_e 8 \cdot 952 = 2 \cdot 1919$
$\log_e 10^2 = 4 \cdot 6052$
$\log_e 895 \cdot 2 = \overline{6 \cdot 7971}$

$\log^e 7 \cdot 686 = 2 \cdot 0394$
$\log_e 10^1 = 2 \cdot 3026$
$\overline{4 \cdot 3420}$
7

$8 \overline{|30 \cdot 3940}$
$\log_e 76 \cdot 86^{\frac{2}{8}} \quad 3 \cdot 7992$

$\log_e 4 \cdot 234 = 1 \cdot 4431$
$\log_e 10^{-3} = \overline{7} \cdot 0922$
$\overline{6 \cdot 5353}$
2

$3 \overline{|11 \cdot 0706}$
$\log_e 0 \cdot 004234^{\frac{2}{3}} = \overline{4} \cdot 3568$

Answer 1046 (b)

A.3

$$Z = \frac{rwl}{\sqrt{r^2 + w^2 l^2}}$$

$$Z^2 = \frac{r^2 w^2 l^2}{r^2 + w^2 l^2}$$

$$Z^2(r^2 + w^2 l^2) = r^2 w^2 l^2$$

$$Z^2 r^2 + Z^2 w^2 l^2 = r^2 w^2 l^2$$

$$Z^2 w^2 l^2 = r^2 w^2 l^2 - Z^2 r^2$$

$$Z^2 w^2 l^2 = r^2 (w^2 l^2 - Z^2)$$

$$r^2 = \frac{Z^2 w^2 l^2}{(w^2 l^2 - Z^2)}$$

$$r = \sqrt{\frac{Z^2 w^2 l^2}{(w^2 l^2 - Z^2)}}$$

$$r = \frac{Zwl}{\sqrt{(wl + Z)(wl - Z)}} \quad Answer$$

A.4 (i) $\quad \dfrac{1}{x} + \dfrac{1}{y} = 1$

(ii) $\quad 2x + 2y = 9$

From (ii)

$$x = \frac{9 - 2y}{2} = 4 \cdot 5 - y$$

Substituting $4.5 - y$ for x in (i)

$$\frac{1}{4.5 - y} + \frac{1}{y} = 1$$

$$\frac{y + 4.5 - y}{y(4.5 - y)} = 1$$

$$y + 4.5 - y = y(4.5 - y)$$

$$4.5 = 4.5y - y^2$$

$$\therefore \quad y^2 - 4.5y + 4.5 = 0$$

$$\left.\begin{array}{l} a = 1 \\ b = -4.5 \\ c = 4.5 \end{array}\right\} \quad y = \frac{4.5 \pm \sqrt{4.5^2 - (4 \times 1 \times 4.5)}}{2 \times 1}$$

$$y = \frac{4.5 \pm \sqrt{20.25 - 18}}{2}$$

$$y = \frac{4.5 \pm \sqrt{2.25}}{2} = \frac{4.5 \pm 1.5}{2}$$

$$\left.\begin{array}{l} \therefore \quad y = 3 \quad \text{when} \quad x = 1.5 \\ \quad\quad y = 1.5 \quad \text{when} \quad x = 3 \end{array}\right\} \quad Answer$$

OR

A.5

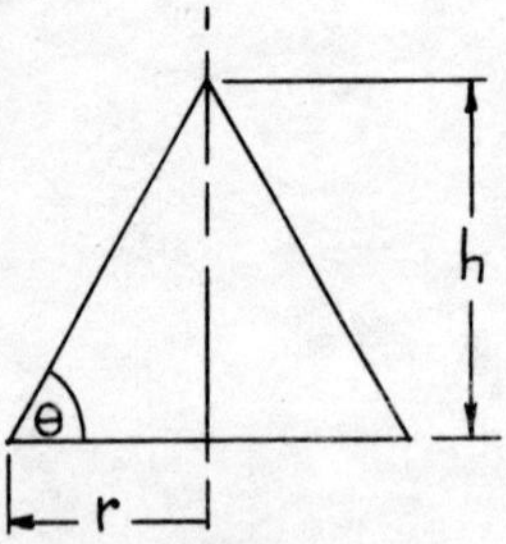

Fig. 3.2

Volume of cone $= \frac{1}{3}\pi r^2 h$

$$\frac{h}{r} = \frac{\text{OPP}}{\text{ADJ}} = \tan\theta$$

$$\therefore \quad h = r \cdot \tan\theta$$

Substituting for h in $\frac{1}{3}\pi r^2 h$

$$Volume = \frac{1}{3}\pi r^2 \cdot r \cdot \tan\theta$$

$$= \frac{\pi r^3 \tan\theta}{3}$$

$$= \frac{3.1416 r^3 \tan\theta}{3} = 1.0472 r^3 \tan\theta$$

r	r^3	$1.0472r^3$	θ	$\tan\theta$	$1.0472r^3 \tan\theta$
1	1	1.0472	0°	0.0000	0
1	1	1.0472	10°	0.1763	0.185
1	1	1.0472	20°	0.3640	0.382
1	1	1.0472	30°	0.5774	0.607
1	1	1.0472	40°	0.8391	0.88
1	1	1.0472	50°	1.1918	1.25
1	1	1.0472	60°	1.7321	1.82
1	1	1.0472	70°	2.7475	2.88
1	1	1.0472	80°	5.6713	5.95

{ Graph to show relationship }
{ between r and θ (0° to 80°) }

Answer

Formula $= 1{\cdot}0472r^3 \tan \theta$

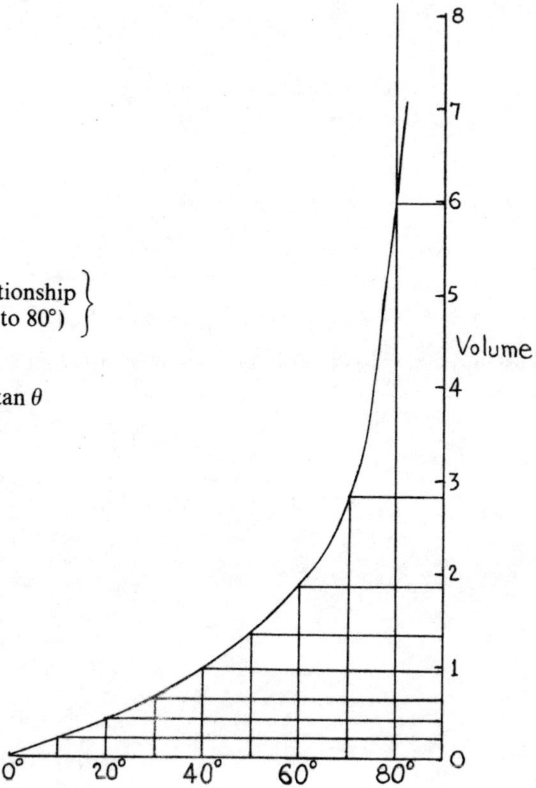

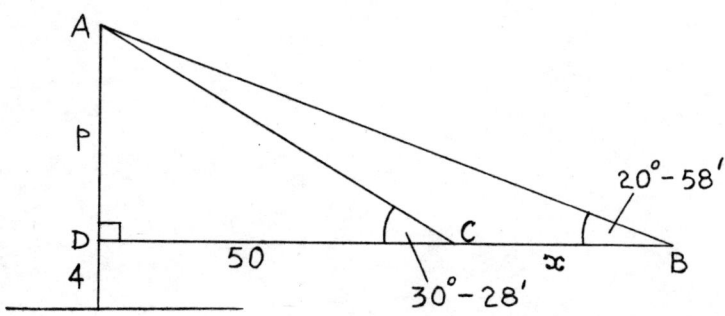

Fig. 3.3

A.6

Fig. 3.4

$$\frac{p}{50} = \tan 30°28'$$

$$\therefore \quad p = 50 \times \tan 30°28'$$

$$= 50 \times 0.5883$$

$$p = 29.415m$$

$$50 + x = p \cdot \cot 20°58'$$

$$= 29.415 \times 2.6092$$

$$\therefore \quad x = 76.7 - 50$$

$$x = 26.7m \quad \text{Distance moved by man (b)}$$

Height of mast above sea-level $= 29.415 + 4 = 33.415m$ (a)

(a) 33.415m ⎫
(b) 26.7m ⎬ *Answer*
        ⎭

A.7

(a)

Fig. 3.5

$$\frac{\frac{4}{3}\pi r^3 - \frac{4}{3}\pi(r-t)^3}{\frac{4}{3}\pi r^3} = \frac{\overset{1}{\frac{4}{3}\pi r^3}}{\frac{4}{3}\pi r^3} - \left(\frac{\frac{4}{3}\pi(r-t)^3}{\frac{4}{3}\pi r^3}\right)$$

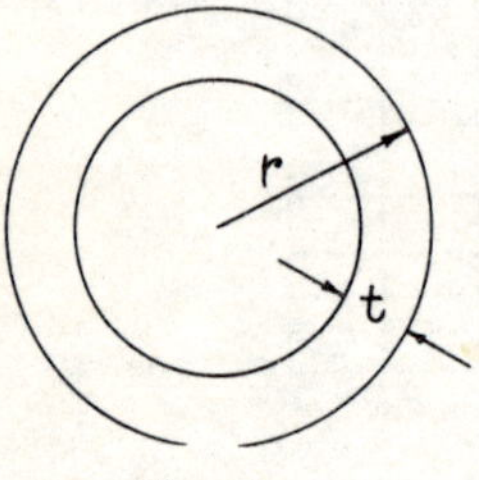

$$= 1 - \left(\frac{(r-t)^3}{r^3}\right)$$

$$= 1 - \left(\frac{(r-t)}{r}\right)^3$$

$$= 1 - \left(\frac{r}{r} - \frac{t}{r}\right)^3$$

$$= 1 - \left(1 - \frac{t}{r}\right)^3 \quad \text{Answer (a)}$$

(b)

$$1 - \left(1 - \frac{t}{r}\right)^3 = 3\frac{t}{r}$$

$$1 - \left(1 - \frac{t}{50t}\right)^3 = 3\frac{t}{50t}$$

$$1 - \left(\frac{50t - t}{50t}\right)^3 = 0.06$$

$$1 - \left(\frac{49t}{50t}\right)^3 = 0.06$$

$$1 - (0.98)^3 = 0.06$$

$$1 - 0.941192 = 0.06$$

$$0.058808 = 0.06$$

$$5.8808\% = 6.00\%$$

$$\therefore \quad \text{Error using } 3\frac{t}{r} = 6.00 - 5.8808 = +0.1192\% \quad \text{Answer (b)}$$

A.8

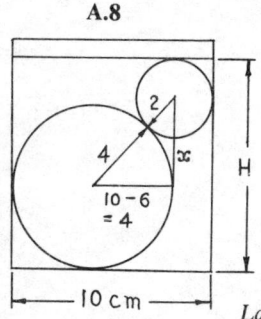

Fig. 3.6

$$x = \sqrt{6^2 - 4^2}$$
$$= \sqrt{36 - 16}$$
$$= \sqrt{20} = 4\cdot472cm$$

∴ Height of water (H)

$$= 4 + 2 + 4\cdot472$$
$$H = 10\cdot472cm$$

Volume of spheres $= \frac{4}{3}\pi r^3$

Large sphere $= \frac{4}{3} \times \frac{22}{7} \times 4^3 = 268\cdot19cm^3$

Small sphere $= \frac{4}{3} \times \frac{22}{7} \times 2^3 = 33\cdot52cm^3$

Volume of cylinder $= \pi r^2 h = \frac{22}{7} \times 5^2 \times 10\cdot472$

(depth of water
     to 'H') $= 824cm^3$

Volume of water $= 824 - (268\cdot19 + 33\cdot52)$
$$= 824 - 301\cdot71$$
$$= 522\cdot29cm^3$$

New height of water (h) $= \dfrac{\text{volume of cylinder} - \text{volume of small sphere}}{\text{Area of bottom of cylinder}}$

$$= \frac{824 - 33\cdot52}{\frac{22}{7} \times 5^2} = \frac{790\cdot48}{78\cdot57} = 10\cdot045cm$$

Volume of water $= 522\cdot29cm^3$
New height of water $= 10\cdot045cm$ $\Big\}$ *Answer*

A.9

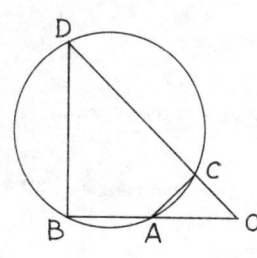

Fig. 3.7

Given: $BD = OB$
To Prove: $AC = OC$
Construction: None

Proof:

$$BD = OB \quad \text{Given}$$

∴ $\triangle OBD$ is isosceles

then

$$B\hat{D}O = B\hat{O}D$$

and

$B\hat{D}O + B\hat{A}C = 180°$ opposite angles of cyclic quadrilateral

$O\hat{A}C + B\hat{A}C = 180°$ OBA is a straight line.

∴ $B\hat{D}O = O\hat{A}C$ External ∠ of cyclic quadrilateral

59

then

$$B\hat{D}O = O\hat{A}C = B\hat{O}D$$

$$\therefore \quad \triangle OAC \text{ is isosceles}$$

and

$$O\hat{A}C = A\hat{O}C$$

then

$$AC = OC \quad Answer$$

TEST PAPER 3

Q.1 Simplify, using indices for (a) and factors for (b)

(a) $$\frac{(h^5 \div h^{-6}) \times h + h^{11}}{(h^{6\cdot5} - h^{5\cdot5})^2 + 2h^{12} - 2h^{11}}$$

(b) $$\frac{(r^2 + 3r - 10)(r^2 + 4r + 4)}{(r^2 + 7r + 10)}$$

Q.2 Evaluate the following using common logarithms

(a) $$(4\cdot829 \log 14)^2$$

(b) Evaluate using Napierian logarithms

$$0\cdot0782^{-0\cdot75}$$

Q.3 Transpose for C in

$$T = 2\pi \cdot \sqrt{\frac{ml^2}{C + \dfrac{ml \cdot MGL}{(L - D)^3}}}$$

Q.4 Solve for x in the following equation, mathematically:

$$x^3 - 6x^2 - 37x + 210 = 0$$

Q.5 A train (A) leaves London at 10.00AM; train (B) leaves at 11.00AM, and (C) at 11.30AM, all travelling at a speed of 20mph.

A slow train (D) leaves at 11.00AM travelling at 10mph and an express (E) leaves at 10.30AM travelling at 40mph. Assuming that some of the trains will pass each other, state which trains are involved and the times of passing.

Q.6 A right-angled triangle is revolved about the vertical axis through an angle of 90°; the vertical and horizontal sides of the triangle being 4 and 3cm respectively. Find the angle between the first and second positions of the hypotenuse.

Q.7 Find the value of a and b in the following:

$$y = a + b \cdot \log_e r$$

When $y = 3; r = 2$ and when $y = 4; r = 5$.

Q.8 A pentagonal shaped rod, whose distance from the centre to one corner is 3-units, has a slice cut off at an angle of 30° to the horizontal, when the rod is lying on its side. Find the ratio between the area of the pentagon and the area of the portion sliced off.

Q.9 Two circles intersect at points P and F, two straight lines AB, CD are drawn through point P, cutting both circles. Prove that

$$\frac{AF}{CF} = \frac{FB}{FD}$$

A.1 (a) $\dfrac{(h^5 \div h^{-6}) \times h + h^{11}}{(h^{6\cdot5} - h^{5\cdot5})^2 + 2h^{12} - 2h^{11}} = \dfrac{h^{12} + h^{11}}{h^{13} - 2h^{12} + h^{11} + 2h^{12} - 2h^{11}}$

$$= \frac{h^{11}(h + 1)}{h^{13} + h^{11} - 2h^{11}}$$

$$= \frac{h^{11}(h + 1)}{h^{11}(h^2 + 1 - 2)}$$

$$= \frac{(h + 1)}{(h^2 - 1)}$$

$$= \frac{(h + 1)}{(h + 1)(h - 1)}$$

$$= \frac{1}{(h - 1)} \quad \textit{Answer}$$

(b) $\dfrac{(r^2 + 3r - 10)(r^2 + 4r + 4)}{(r^2 + 7r + 10)} = \dfrac{(r + 5)(r - 2)(r + 2)(r + 2)}{(r + 5)(r + 2)}$

$$= (r + 2)(r - 2)$$

$$= (r^2 - 2^2) \quad \textit{Answer}$$

A.2 (a) $(4\cdot829 \,.\, \log 14)^2 = (4\cdot829 \times 1\cdot1461)^2$

$$= 2(\log 4\cdot829 + \log 1\cdot1461)^2$$

$$= 2(0\cdot6838 + 0\cdot0592)^2$$

$$= 2 \times 0\cdot7430$$

$$= 1\cdot4860$$

$$\text{Antilogging} = 30\cdot62 \quad \textit{Answer}$$

(b) $0\cdot0782^{-0.75} = -0\cdot75 \times \log_e 0\cdot0782$

$$= -0\cdot75 \times \bar{3}\cdot4515$$

$$= \frac{-3 \times \bar{3}\cdot4515}{4}$$

$$= \frac{-3 \times -2\cdot5485}{4}$$

$$= \frac{7\cdot6455}{4}$$

$$= 1\cdot9114$$

$$\text{Antilogging} = 6\cdot763 \quad \textit{Answer}$$

A.3

$$T = 2\pi . \sqrt{\dfrac{ml^2}{C + \dfrac{ml . MGL}{(L - D)^3}}}$$

$$T^2 = 4\pi^2 \times \dfrac{ml^2}{C + \dfrac{ml . MGL}{(L - D)^3}}$$

$$T^2 \left(C + \dfrac{ml . MGL}{(L - D)^3} \right) = 4\pi^2 \times ml^2$$

$$C + \dfrac{ml . MGL}{(L - D)^3} = \dfrac{4\pi^2 . ml^2}{T^2}$$

$$\therefore \quad C = \dfrac{4\pi^2 ml^2}{T^2} - \dfrac{ml . MGL}{(L - D)^3}$$

$$C = ml \left(\dfrac{4\pi^2 l}{T^2} - \dfrac{MGL}{(L - D)^3} \right) \quad Answer$$

A.4 $x^3 - 6x^2 - 37x + 210 = 0$

x	x^2	x^3	$x^3 - 6x^2 - 37x + 210$	y
-1	1	-1	$-1 - 6 + 37 + 210$	240
$+1$	1	1	$1 - 6 - 37 + 210$	168
-2	4	-8	$-8 - 24 + 74 + 210$	252
$+2$	4	8	$8 - 24 - 74 + 210$	120
-3	9	-27	$-27 - 54 + 111 + 210$	240
$+3$	9	27	$27 - 54 - 111 + 210$	72
-4	16	-64	$-64 - 96 + 148 + 210$	198
$+4$	16	64	$64 - 96 - 148 + 210$	30
-5	25	-125	$-125 - 150 + 185 + 210$	120
$+5$	25	125	$125 - 150 - 185 + 210$	0 ←

$\therefore$ If $x = 5$ is a root, then $(x - 5)$ is a factor.

$$x - 5)x^3 - 6x^2 - 37x + 210(x^2 - x - 42$$

$\therefore \quad (x - 5)(x^2 - x - 42) = x^3 - 6x^2 - 37x + 210.$

Factorizing

$$(x^2 - x - 42) = 0$$

$$x^2 - 7x + 6x - 42 = 0$$

$$x(x - 7) + 6(x - 7) = 0$$

$$(x + 6)(x - 7) = 0$$

$$\therefore \quad x = -6 \text{ or } 7$$

and

$$x = -6, 5 \text{ or } 7 \quad Answer$$

A.5

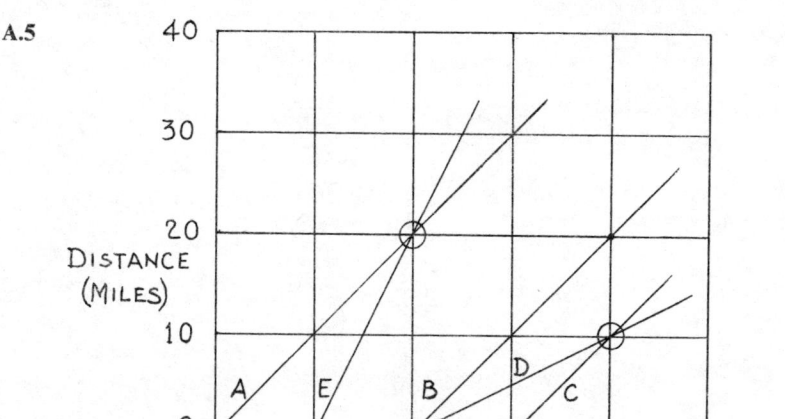

Fig. 3.8

Express (*E*) passes (*A*) at 1100 hours ⎫
(*C*) passes (*D*) at 1200 hours ⎭ *Answer*

A.6

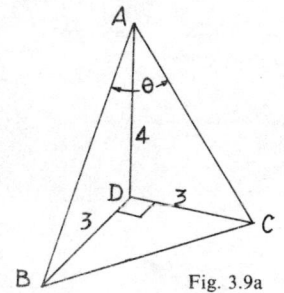

Fig. 3.9a

Sides *AB* and *AC*

$$= \sqrt{4^2 + 3^2}$$
$$= \sqrt{16 + 9}$$
$$= \sqrt{25}$$
$$= 5cm$$

Side $BC = \sqrt{3^2 + 3^2}$
$$= \sqrt{9 + 9}$$
$$= \sqrt{18} = 4.243cm$$

Using cosine rule:

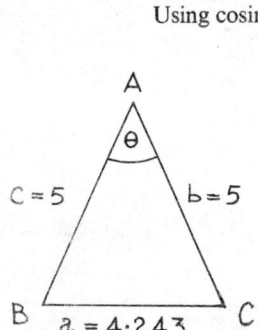

Fig. 3.9b

$$\cos A = \frac{b^2 + c^2 - a^2}{2bc}$$

$$\cos A = \frac{5^2 + 5^2 - 4.243^2}{2 \times 5 \times 5}$$

$$= \frac{25 + 25 - 18}{50}$$

$$= \frac{50 - 18}{50}$$

$$\cos A = \frac{32}{50} = 0.64$$

$$\therefore \quad \angle A = 50°12' \quad Answer$$

63

A.7

$$y = a + b . \log_e r$$

When $y = 3 ; r = 2$ and when $y = 4 ; r = 5$

    ① $3 = a + b . \log_e 2$

    ② $4 = a + b . \log_e 5$

Subtracting ① from ②

    ② $4 = a + b \times 1.6094$

    ① $3 = a + b \times 0.6931$

    $\overline{1 = \qquad b \times 0.9163}$

$$\therefore \quad b = \frac{1}{0.9163} = 1.0913$$

Substituting value of $b = 1.0913$ in ① above

$$3 = a + 1.0913 \times 0.6931$$
$$\therefore \quad a = 3 - 0.75638$$
$$a = 2.24362$$

$$a = 2.24362 ; b = 1.0913 \quad \textit{Answer}$$

A.8

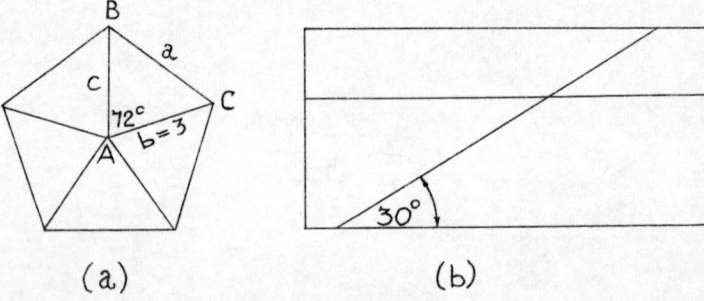

Fig. 3.10

$$\text{Area of } \triangle ABC = \tfrac{1}{2} . bc . \sin \theta$$
$$= \tfrac{1}{2} \times 3^2 \times \sin 72°$$
$$= \tfrac{9}{2} \times 0.9511$$
$$= 4.28 \text{ units}^2$$
$$\textit{Area of pentagon} = 4.28 \times 5$$
$$= 21.4 \text{ units}^2$$

$$Area \ of \ slice = \frac{21 \cdot 4}{\sin 30°}$$

$$= \frac{21 \cdot 4}{0 \cdot 5}$$

$$= 42 \cdot 8 \ units^2$$

Ratio of pentagon to slice $= 21 \cdot 4 : 42 \cdot 8$

$$= 1 : 2$$

Ratio $1 : 2$ *Answer*

A.9

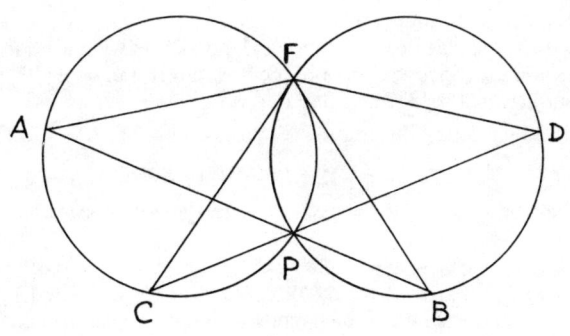

Fig. 3.11

Given: Intersecting circles and lines AB and CD

To Prove:
$$\frac{AF}{CF} = \frac{FB}{FD}$$

Construction: Join AF, CF, FD, FB

Proof:
$$\angle A = \angle C \quad \text{On same arc } FP$$
$$\angle D = \angle B \quad \text{On same arc } FP$$
$$\therefore \quad C\widehat{F}D = A\widehat{F}B$$

Then $\triangle$'s AFB and CFD are similar

$$\therefore \quad \frac{AF}{CF} = \frac{FB}{FD} \quad \textit{Answer}$$

TEST PAPER 4

Q.1 Simplify by removing all root signs and making all indices positive:

$$\sqrt[3]{\frac{(x+1)-(x+1)^{-0.5}}{(x+1)+(x+1)}}$$

Q.2 Evaluate using Napierian logarithms only:

(a)
$$(0.5472)^{-1.423}$$

(b)
$$\frac{6.243^{0.5432}}{8.274}$$

Q.3 Given that

$$1-\left(\frac{V_1}{V_2}\right)^{\gamma-1}=0.548$$

and $P_1V_1^\gamma = P_2V_2^\gamma$. Where $P_1 = 92$; $V_1 = 4.2$; $V_2 = 21$; Find P_2.

Q.4 (a) Establish the formula for the solution of a quadratic equation.
(b) A closed rectangular box is 6 times as long as it is wide, with a height of 8cm. Find an expression for the surface area of the box in terms of its width.
(c) If the surface area is 1372cm^2, find the width.

Q.5 (a) Prove by plotting the graph or otherwise, that

$$(x-2)^2+(y-3)^2=25$$

(b) Also find the radius and the points where the equation $4y = 3x + 6$ cuts the circle.

Q.6 A man, whose eyes were at a height of 173cm above the ground, sighted a flagpole at an elevation of 27°. The man then walks 10 metres towards the flagpole when the elevation is 55°. Find the height of the flagpole and the original distance.

Q.7 A circle of 25mm radius with centre 0, has a tangent (AB) to the circle, drawn horizontally. Point B is at the point of contact, and a perpendicular is dropped from O to B. A line is drawn from A at 20° to the horizontal. At the point where OB cuts the circle a line is drawn to the point where AO cuts the circle. Calculate the ratio of the circumference of the circle to the perimeter of the triangle OCB. Also calculate the perimeter of triangle ABC.

Q.8 (a) Develop the formula for the total surface area of a cone.
(b) Find the area if the vertical height is 7cm and has a diameter of 6cm.

Q.9 In a triangle ABC, $AB = AC$. If X and Y are points on AB and AC respectively, such that $AX = AY$. Prove $XC = YB$. If XC and YB intersect at P. Prove $PC = PB$.

A.1
$$\sqrt[3]{\frac{(x+1)-(x+1)^{-0.5}}{(x+1)+(x+1)}}=\left(\frac{(x+1)-(x+1)^{-0.5}}{(x+1)+(x+1)}\right)^{\frac{1}{3}}$$

$$=\left(\frac{(x+1)}{2(x+1)}-\frac{1}{2(x+1)(x+1)^{0.5}}\right)^{\frac{1}{3}}$$

$$=\left(\frac{1}{2}-\frac{1}{2(x+1)^{\frac{3}{2}}}\right)^{\frac{1}{3}}$$

$$=\left[\frac{1}{2}\left(1-\frac{1}{(x+1)^{\frac{3}{2}}}\right)\right]^{\frac{1}{3}} \quad Answer$$

A.2 (a) $(0.5472)^{-1.423}$

$$-1.423 \times \log_e 0.5472$$

$$-1.423 \times \bar{1}.3971$$

$$-1.423 \times -0.6029$$

Using loglogs $\quad \log_e -0.6029 = \bar{1}.4940$
$\log_e -1.423 = 0.3528$
$$\overline{\bar{1}.8468}$$
Antilogging 1st time $\quad \bar{3}.6974$
$$\overline{2.1494}$$
$$= 0.8580$$
Antilogging 2nd time $= 2.3580$ $\quad$ *Answer* (a)

(b)
$$\left(\frac{6.243}{8.274}\right)^{0.5432}$$

$$\log_e 6.243 = 1.8315$$
$$\log_e 8.274 = 2.1131$$
$$\overline{\bar{1}.7184}$$

$$= 0.5432 \times \bar{1}.7184$$

$$0.5432 \times -0.2816$$

Using loglogs $\quad \log_e -0.2816 = \quad \bar{2}.7327$
$\log_e 0.5432 = \quad \bar{1}.3897$
Antilogging 1st time $= \quad \overline{\bar{2}.1224}$
$$\bar{3}.6974$$
$$\overline{0.4250}$$
$$- \quad -0.1530$$
$$-1+1$$
Antilogging 2nd time $= \quad \overline{\bar{1}.8470}$
$$\bar{3}.6974$$
$$\overline{2.1496}$$

$$\therefore \quad Answer \ (b) = 0.8582$$

A.3
$$1 - \left(\frac{V_1}{V_2}\right)^{\gamma-1} = 0.548$$

$$1 - \left(\frac{4.2}{21.0}\right)^{\gamma-1} = 0.548$$

Where $P_1 = 92$
$P_2 = ?$
$V_1 = 4.2$
$V_2 = 21$

$$1 - (0.2)^{\gamma-1} = 0.548$$

$$1 - 0.548 = 0.2^{\gamma-1}$$

$$\log 0.452 = (\gamma - 1)\log 0.2$$

$$\bar{1}.6551 = (\gamma - 1)\bar{1}.3010$$

$$(\gamma - 1) = \frac{\bar{1}.6551}{\bar{1}.3010}$$

$$(\gamma - 1) = \frac{-0 \cdot 3449}{-0 \cdot 6990} = 0 \cdot 4934$$

$$\gamma = 1 + 0 \cdot 4934$$

$$\gamma = 1 \cdot 4934 \quad Answer$$

$$P_1 V_1^\gamma = P_2 V_2^\gamma$$

$$\therefore \quad P_2 = P_1 \left(\frac{V_1}{V_2}\right)^{1 \cdot 4934}$$

$$P_2 = 92 \cdot \left(\frac{4 \cdot 2}{21}\right)^{1 \cdot 4934}$$

$$\log P_2 = \log 92 + 1 \cdot 4934 \log 0 \cdot 2$$

$$= 1 \cdot 9638 + 1 \cdot 4934 \times \bar{1} \cdot 3010$$

$$= 1 \cdot 9638 + 1 \cdot 4934 \times -0 \cdot 699$$

$$= 1 \cdot 9638 - 1 \cdot 0439$$

$$= 0 \cdot 9199$$

Antilogging $P_2 = 8 \cdot 316 \quad Answer$

A.4 (a)

$$ax^2 + bx + c = 0$$

$$x^2 + \frac{b}{a}x + \frac{c}{a} = 0$$

$$x^2 + \frac{b}{a}x = -\frac{c}{a}$$

$$x^2 + \frac{b}{a}x + \left(\frac{b}{2a}\right)^2 = -\frac{c}{a} + \left(\frac{b}{2a}\right)^2$$

$$\left(x + \frac{b}{2a}\right)^2 = -\frac{c}{a} + \frac{b^2}{4a^2}$$

$$\left(x + \frac{b}{2a}\right)^2 = \frac{-4ac + b^2}{4a^2}$$

$$x + \frac{b}{2a} = \pm\sqrt{\frac{-4ac + b^2}{4a^2}}$$

$$x + \frac{b}{2a} = \frac{\pm\sqrt{b^2 - 4ac}}{2a}$$

$$x = -\frac{b}{2a} \frac{\pm\sqrt{b^2 - 4ac}}{2a}$$

$$x = \frac{-b \pm \sqrt{b^2 - 4ac}}{2a} \quad Answer \ (a)$$

(b) *Surface Area* $= 2[(6W \times W) + (8 \times W) + (8 \times 6W)]$
$$= 2[6W^2 + 8W + 48W]$$
$$= 2[6W^2 + 56W]$$
$$= 12W^2 + 112W \quad Answer \ (b)$$

(c) $\qquad 1372 = 12W^2 + 112W$
$$686 = 6W^2 + 56W$$
$$6W^2 + 56W - 686 = 0$$
$$\left. \begin{array}{l} a = 6 \\ b = 56 \\ c = -686 \end{array} \right\}$$

Substitute above values in formula:

$$W = \frac{-b \pm \sqrt{b^2 - 4ac}}{2a}$$

$$W = \frac{-56 \pm \sqrt{56^2 - (4 \times 6 \times -686)}}{2 \times 6}$$

$$W = \frac{-56 \pm \sqrt{3136 + 16\,464}}{12}$$

$$W = \frac{-56 \pm \sqrt{19\,600}}{12}$$

$$W = \frac{-56 \pm 140}{12}$$

$$W = \frac{-56 - 140}{12} = \frac{-196}{12} = -16.33$$

$$\therefore \quad W = \frac{-56 + 140}{12} = \frac{84}{12} = 7 \quad Answer \ (c)$$

A.5 (a)

From $(x - 2)^2 + (y - 3)^2 = 25$

Centre Circle: $x = 2$; $y = 3$ and $r = 5$

(b) $\qquad\qquad\qquad 4y = 3x + 6$
$$y = \tfrac{3}{4}x + 1\tfrac{1}{2}$$

x	-2	2	4
$\tfrac{3}{4}x$ $+1\tfrac{1}{2}$	$-1\tfrac{1}{2}$ $1\tfrac{1}{2}$	$1\tfrac{1}{2}$ $1\tfrac{1}{2}$	3 $1\tfrac{1}{2}$
y	0	3	$4\tfrac{1}{2}$

69

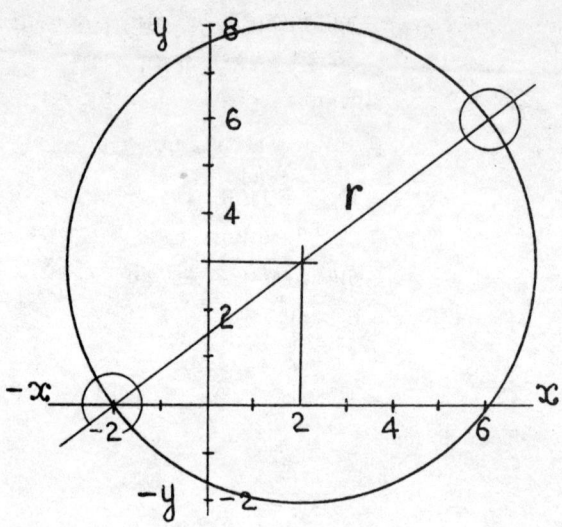

Fig. 3.12

Graph cuts circle at $\left.\begin{array}{l} x = 6; y = 6 \\ x = -2; y = 0 \end{array}\right\}$ (b)

radius = 5

Centre of circle $(2, 3)$ (α) *Answer*

A.6

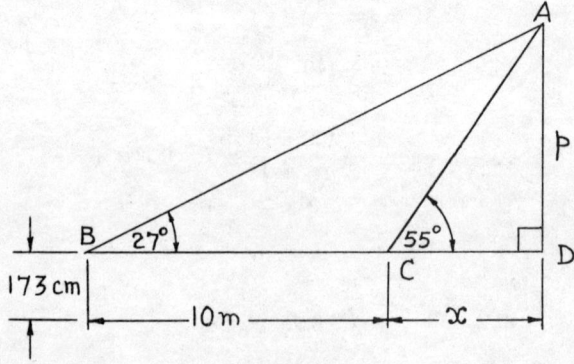

Fig. 3.13

$$x = p \cdot \cot 55°$$
$$= 0.7002p$$
$$10 + x = p \cdot \cot 27°$$
$$= 1.9626p$$

$\therefore$ Subtracting, we get:

$$10 = p(1{\cdot}9626 - 0{\cdot}7002) = 1{\cdot}2624p$$

$$\therefore \quad p = \frac{10}{1{\cdot}2624} = 7{\cdot}925m$$

$$x = p \,.\, \cot 55° = 7{\cdot}925 \times 0{\cdot}7002$$

$$x = 5{\cdot}5491m$$

Original Distance $= x + 10 = 5{\cdot}5491 + 10$

$$= 15{\cdot}5491m$$

Answer $\left.\begin{array}{l} 15{\cdot}5491m \\ 7{\cdot}925m\,. \end{array}\right\}$

A.7 (a)

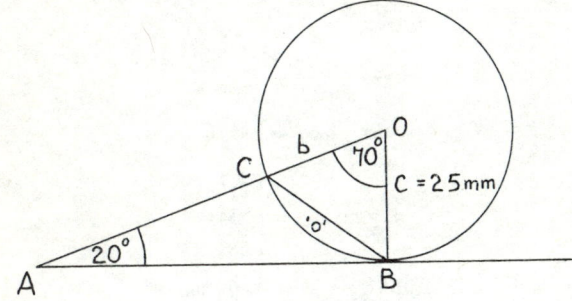

Fig. 3.14

$\triangle AOB$ is right-angled

$$\therefore \quad \angle AOB = 90° - 20°$$

$$= 70°$$

and $\angle$'s OCB and $OBC = 55°$

Ratio $2\pi r : {}^\circ O' + c + b$

$$\frac{{}^\circ O'/2}{OC} = \sin 35°$$

$$\therefore \quad \frac{{}^\circ O'}{2} = OC \,.\, \sin 35°$$

$${}^\circ O' = 2 \times OC \times \sin 35°$$

$$= 2 \times 25 \times 0{\cdot}5736$$

$$\therefore \quad {}^\circ O' = 28{\cdot}68mm$$

Perimeter of $\triangle OCB = {}^\circ O' + c + b$

$$= 28{\cdot}68 + 25 + 25$$

$$= 78{\cdot}68mm$$

71

Circumference of circle $= 2\pi r$

$$= 2 \times 3.1416 \times 25$$

$$= 157.08mm$$

Ratio   157.08 : 78.68   log 157.08 = 2.1961
            2 : 1   (a)   log 78.68 = 1.8958
                                              ‾‾‾‾‾‾
                                              0.3003

Antilogging $= 1.996$ say 2

(b)

$$\frac{a}{\sin A} = \frac{b}{\sin B}$$

$$\frac{28.68}{\sin 20°} = \frac{b}{\sin 35°}$$

$$\therefore \quad b = \frac{28.68 \times \sin 35°}{\sin 20°}$$

$$= \quad \begin{array}{l} \log 28.68 = 1.4576 \\ \log \sin 35° = \bar{1}.7586 \\ \hline 1.2162 \\ \log \sin 20° = \bar{1}.5341 \\ \hline 1.6821 \end{array}$$

Antilogging $b = 48.09mm$

$$\frac{a}{\sin A} = \frac{c}{\sin C}$$

$$c = \frac{a \cdot \sin C}{\sin A}$$

$$c = \frac{28.68 \times \sin 125°}{\sin 20°}$$

$$= \quad \begin{array}{l} \log 28.68 = 1.4576 \\ \log \sin 125° = \bar{1}.9134 \\ \hline 1.3710 \\ \log \sin 20° = \bar{1}.5341 \\ \hline 1.8369 \end{array}$$

Antilogging $c = 68.69mm$

Perimeter $= a + b + c$

$$= 28.68 + 48.09 + 68.69$$

$$= 145.46mm \quad (b) \; \Big\}$$
$$Ratio = 2 : 1 \quad (a) \; \Big\} \quad Answer$$

A.8 (a)

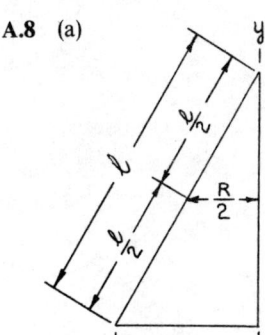

Fig. 3.15(a)

By Theorem of Pappus:

Area of curved surface of cone

= *Generated area*

= Length of arm × distance moved by centroid

$$= l \times 2\pi \frac{R}{2}$$

$$= \pi R l$$

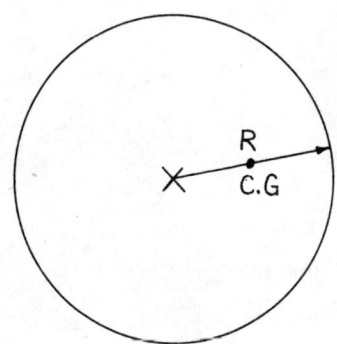

Fig. 3.15(b)

Area of base

= *Generated area*

= Length of arm × distance moved by C.G.

$$= R \times 2\pi \frac{R}{2}$$

$$= \pi R^2$$

(b)

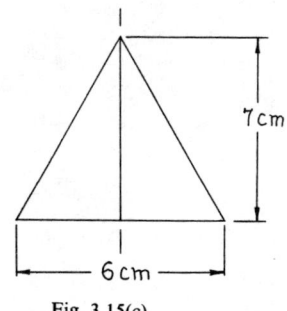

Fig. 3.15(c)

Total area of cone $= \pi R l + \pi R^2$

$$= \pi R(l + R)$$

Slant height $= \sqrt{7^2 + 3^2}$

$$= \sqrt{58}$$

$$= 7.616 cm$$

Total Surface Area $= \pi R(l + R)$

$$= \tfrac{22}{7} \times 3(7.616 + 3)$$

$$= \tfrac{66}{7} \times 10.616$$

$$= 100.089 cm^2 \quad Answer$$

A.9

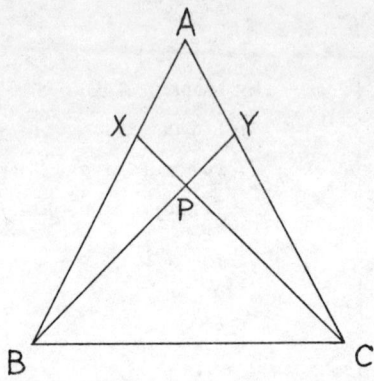

Fig. 3.16

(i) In $\triangle$'s AXC and AYB

$$AB = AC \qquad \text{Given}$$
$$\angle A \qquad \text{is common}$$
$$AX = AY \qquad \text{given}$$

$\therefore$ $\triangle$'s are congruent SAS

$\therefore$ $XC = YB$ (i)

(ii) $A\widehat{X}C = X\widehat{B}C + X\widehat{C}B$ (External $\angle$ + 2 internal opposite $\angle$'s)

$\quad A\widehat{Y}B = Y\widehat{B}C + Y\widehat{C}B$ (External $\angle$ + 2 internal opposite $\angle$'s)

$\quad A\widehat{X}C = A\widehat{Y}B$

$\therefore$ $X\widehat{B}C + X\widehat{C}B = Y\widehat{B}C + Y\widehat{C}B$

$\qquad X\widehat{B}C = Y\widehat{C}B$

$\therefore$ $X\widehat{C}B = Y\widehat{B}C$

Then $\triangle BPC$ is isosceles (Equal sides opposite equal angles)

$$\therefore \quad BP = PC \quad \text{(ii)}$$

TEST PAPER 5

Q.1 Simplify:

(a)
$$\frac{[x(x^2y^3)]^{\frac{2}{3}}}{x^3y^3}$$

(b)
$$\left[x^{\frac{2}{3}} \cdot \left(\frac{1}{xy} \right)^{\frac{1}{3}} \times \left(\frac{xy^{-\frac{1}{3}}}{xy^{-\frac{2}{3}}} \right)^{-\frac{1}{2}} \right]^{\frac{2}{3}}$$

Q.2 Without using log tables, evaluate:

(a) $$3\log_{10} 2^2 + \log_{10} 4^2 - \tfrac{1}{2}\log_{10} 6^4$$

(b) Given $x = e^{P-CV}$ and $y = e^{P+CV}$; find the value of

(i) xy;   (ii) $\dfrac{x}{y^3}$;   (iii) $\dfrac{y^3}{x}$

Q.3 Solve for W in the following equation:

$$Z = \sqrt{R^2 + \left(WL - \frac{1}{WC}\right)^2}$$

Q.4 (a) Explain in simple terms what is meant by the equation $2(x + y) = 22$ with respect to a rectangle whose breadth is x and length y.

(b) A rectangle has a diagonal 5cm long. Find the diameter in centimetres of a circle having the same area, given that the length of the rectangle is twice its breadth.

Q.5

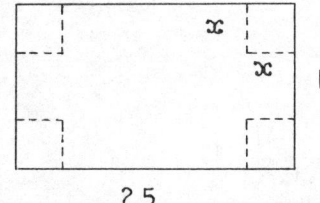

12

25

Fig. 3.17

(a) From the information given in the diagram, make up a formula in terms of x for the volume of a box.

(b) Given the following values for x; 0, 1, 2, 3, 4, 5, 6 construct a graph and find the value of x which gives the maximum volume if the above figure is shaped in the form of a box.

Q.6 A point C is 170m due north of point A, and a point B is lying so that it is 80m from point C and 150m from point A. A flagpole m is at point B and is at an elevation of $20°$ from A.

(a) Find the height of the pole.

(b) Prove that triangle ABC is right-angled.

(c) Find the angle of m from a point W which is lying on line AC and is due west of B.

Q.7 A concrete pyramid of 24 metres perpendicular height and 12m square base was put in position as a navigational mark at the entrance to a river. Calculate the volume of the pyramid using Simpson's Rule, and not less than five ordinates.

Q.8 A piece of wire is 4 units long. A certain length is cut off and formed into the circumference of a circle. Prove that the area of the circle is equal to $\dfrac{x^2}{4\pi}$. The remaining wire is formed into a square. Prove that the area of the square is equal to $1 - \dfrac{x}{2} + \dfrac{x^2}{16}$.

75

Q.9

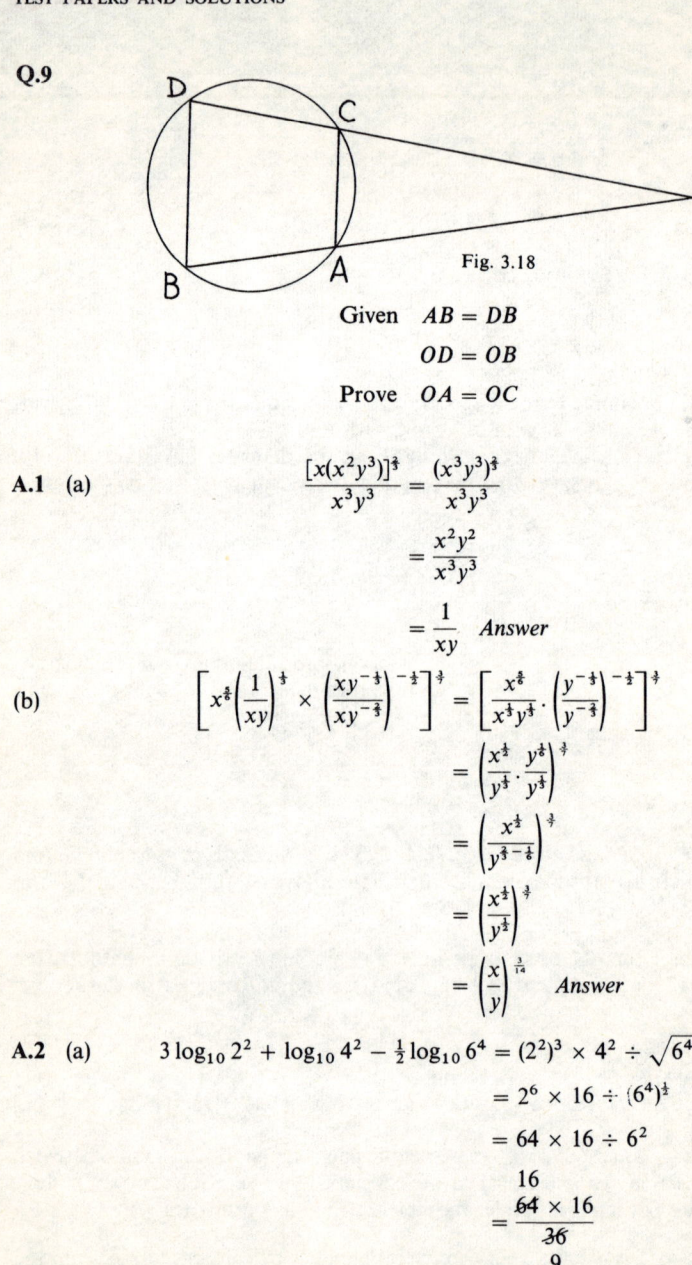

Fig. 3.18

$$\text{Given} \quad AB = DB$$
$$OD = OB$$
$$\text{Prove} \quad OA = OC$$

A.1 (a)

$$\frac{[x(x^2y^3)]^{\frac{2}{3}}}{x^3y^3} = \frac{(x^3y^3)^{\frac{2}{3}}}{x^3y^3}$$

$$= \frac{x^2y^2}{x^3y^3}$$

$$= \frac{1}{xy} \quad Answer$$

(b)

$$\left[x^{\frac{2}{5}}\left(\frac{1}{xy}\right)^{\frac{1}{3}} \times \left(\frac{xy^{-\frac{1}{3}}}{xy^{-\frac{2}{3}}}\right)^{-\frac{1}{2}}\right]^{\frac{2}{3}} = \left[\frac{x^{\frac{2}{5}}}{x^{\frac{1}{3}}y^{\frac{1}{3}}} \cdot \left(\frac{y^{-\frac{1}{3}}}{y^{-\frac{2}{3}}}\right)^{-\frac{1}{2}}\right]^{\frac{2}{3}}$$

$$= \left(\frac{x^{\frac{1}{5}}}{y^{\frac{1}{3}}} \cdot \frac{y^{\frac{1}{6}}}{y^{\frac{1}{3}}}\right)^{\frac{2}{3}}$$

$$= \left(\frac{x^{\frac{1}{5}}}{y^{\frac{1}{3}-\frac{1}{6}}}\right)^{\frac{2}{3}}$$

$$= \left(\frac{x^{\frac{1}{5}}}{y^{\frac{1}{2}}}\right)^{\frac{2}{3}}$$

$$= \left(\frac{x}{y}\right)^{\frac{3}{14}} \quad Answer$$

A.2 (a)

$$3\log_{10} 2^2 + \log_{10} 4^2 - \tfrac{1}{2}\log_{10} 6^4 = (2^2)^3 \times 4^2 \div \sqrt{6^4}$$

$$= 2^6 \times 16 \div (6^4)^{\frac{1}{2}}$$

$$= 64 \times 16 \div 6^2$$

$$= \frac{\overset{16}{\cancel{64}} \times 16}{\underset{9}{\cancel{36}}}$$

$$= \frac{256}{9}$$

$$= 28\tfrac{4}{9} \quad Answer$$

(b) $\qquad x = e^{P-CV} \quad \text{and} \quad y = e^{P+CV}$

(i) $\qquad xy = e^{P-CV} \times e^{P+CV}$

$$= e^{P-CV+P+CV}$$

$$xy = e^{2P} \quad \textit{Answer (i)}$$

(ii) $\qquad \dfrac{x}{y^3} = \dfrac{e^{P-CV}}{e^{(P+CV)3}}$

$$= e^{P-CV-3P-3CV}$$

$$= e^{-2P-4CV}$$

$$\dfrac{x}{y^3} = \dfrac{1}{e^{2P+4CV}} \quad \textit{Answer (ii)}$$

(iii) $\qquad \dfrac{y^3}{x} = \dfrac{e^{(P+CV)3}}{e^{(P-CV)}}$

$$= \dfrac{e^{3P+3CV}}{e^{P-CV}}$$

$$= e^{3P+3CV-P+CV}$$

$$\dfrac{y^3}{x} = e^{2P+4CV} \quad \textit{Answer (iii)}$$

A.3 $\qquad\qquad Z = \sqrt{R^2 + \left(WL - \dfrac{1}{WC}\right)^2}$

$$Z^2 = R^2 + \left(WL - \dfrac{1}{WC}\right)^2$$

$$Z^2 - R^2 = \left(WL - \dfrac{1}{WC}\right)^2$$

$$\sqrt{Z^2 - R^2} = \left(WL - \dfrac{1}{WC}\right)$$

$$W\sqrt{Z^2 - R^2} = W^2L - \dfrac{1}{C}$$

$$W^2L - W\sqrt{Z^2 - R^2} - \dfrac{1}{C} = 0$$

$$\left.\begin{array}{l} a = L \\ b = -\sqrt{Z^2 - R^2} \\ c = -\dfrac{1}{C} \end{array}\right\}$$

$$W = \frac{\sqrt{Z^2 - R^2} \pm \sqrt{Z^2 - R^2 - \left(4 \times L \times -\dfrac{1}{C}\right)}}{2 \times L}$$

$$W = \frac{\sqrt{Z^2 - R^2} \pm \sqrt{Z^2 - R^2 + \dfrac{4L}{C}}}{2L}$$

$$W = \frac{1}{2L}\left(\sqrt{Z^2 - R^2} \pm \sqrt{Z^2 - R^2 + \frac{4L}{C}}\right) \quad Answer$$

A.4 (a)

$$2(x + y) = 22$$

$$2x + 2y = 22 = \text{Perimeter of rectangle.}$$

(b)

$\therefore$ By Pythagoras

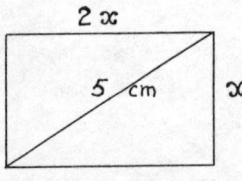

Fig. 3.19

$$(2x)^2 + x^2 = 5^2$$

$$4x^2 + x^2 = 25$$

$$5x^2 = 25$$

$$x = \sqrt{\frac{25}{5}}$$

$$x = \sqrt{5} = 2\cdot236\text{cm}$$

$\therefore$ Breadth of rectangle $= 2\cdot236\text{cm}$

Length of rectangle $= 4\cdot472\text{cm}$

Area of rectangle $= 2\cdot236 \times 4\cdot472$

$$= 10cm^2$$

Area of circle $= \dfrac{\pi}{4}D^2$

$$\therefore \quad \frac{\pi}{4}D^2 = 10$$

$$\pi D^2 = 40$$

$$D^2 = \frac{40}{\pi} \quad \text{Where } \pi = \tfrac{22}{7}$$

$$D = \sqrt{40 \times \tfrac{7}{22}}$$

$$D = \sqrt{12\cdot72} = 3\cdot567\text{cm}$$

$\therefore$ Diameter of circle $= 3\cdot567\text{cm}$ *Answer*

A.5 (a)

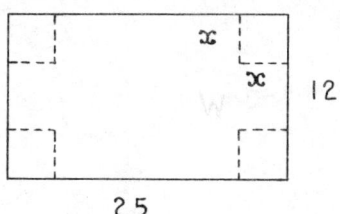

Fig. 3.20

$$\text{Volume} = (25 - 2x)(12 - 2x)x$$
$$= 4x^3 - 74x^2 + 300x$$

x	x^2	x^3	$4x^3 - 74x^2 + 300x$	y
0	0	0	0 − 0 + 0	0
1	1	1	4 − 74 + 300	230
2	4	8	32 − 296 + 600	336
3	9	27	108 − 666 + 900	342
4	16	64	256 − 1184 + 1200	272
5	25	125	500 − 1850 + 1500	150
6	36	216	864 − 2664 + 1800	0

$x = 2.6$ approximately gives maximum volume.

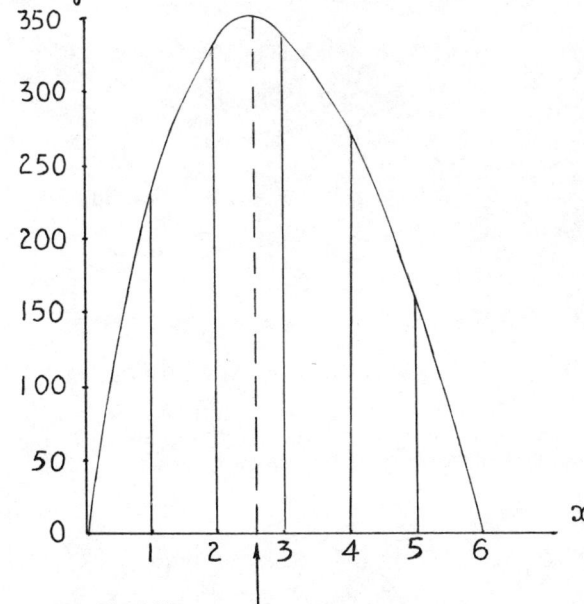

Fig. 3.21

$x = 2.6$ approximately *Answer*

79

A.6 (a)

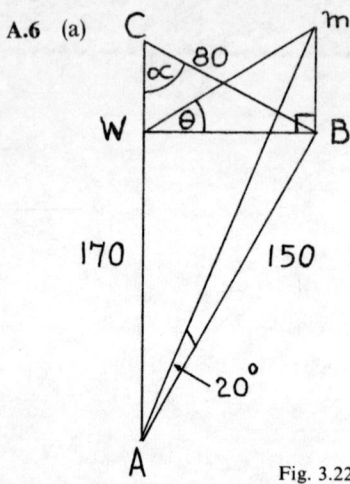

Fig. 3.22

Height of pole (mb)

$$= \frac{mB}{AB} = \tan 20°$$

$$\therefore \quad mB = AB \cdot \tan 20°$$

$$= 150 \times 0.3640$$

$$= 54.6\text{m} \quad Answer \ (a)$$

(b) By Pythagoras

$$b^2 = a^2 + c^2$$

$$170^2 = 80^2 + 150^2$$

$$\sqrt{28\,900} = \sqrt{28\,900}$$

Answer (b) $170 = 170$ $\therefore$ $\triangle ABC$ must be right $\angle$'d.

(c)

$$\angle C = \frac{150}{170} = \sin \alpha = 0.8824 = 61°56'$$

$$WB = CB \cdot \sin \alpha$$

$$= 80 \times 0.8824 = 70.592m$$

$$\angle W = \frac{mB}{WB} = \frac{54.6}{70.592} = \tan \theta = 0.7734$$

$$\therefore \quad \angle W = 37°43' \quad Answer \ (c)$$

A.7

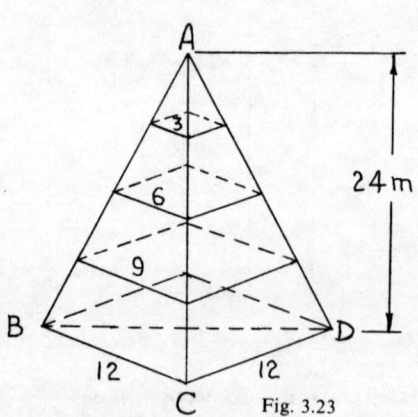

Fig. 3.23

Using Simpson's 1st Rule:

$$\frac{h}{3}(1 + 4 + 2 + 4 + 1)$$

Where $h = \frac{24}{4} = 6m$

Areas	S.M's	Products
0	1	0
9	4	36
36	2	72
81	4	324
144	1	144
		576

$$\frac{h}{3} \times 576 = \frac{\overset{2}{6}}{3} \times 576 = 1152$$

$$\therefore \quad \text{Volume} = 1152m^3 \quad Answer$$

A.8 (a) Circumference of circle $= \pi d$

$$\therefore \quad \pi d = x \quad \text{and} \quad d = \frac{x}{\pi}$$

Area of circle $= \frac{\pi}{4}d^2$

Substituting for d:

$$\frac{\pi}{4}d^2 = \frac{\pi}{4}\left(\frac{x}{\pi}\right)^2$$

$$= \frac{\pi}{4} \cdot \frac{x^2}{\pi^2}$$

$$\textit{Area of circle} = \frac{x^2}{4\pi} \quad Answer \ (a)$$

(b) Perimeter of square $= (4 - x)$

$$\text{Area of square} = \left(\frac{(4 - x)}{4}\right)^2$$

$$= \left(\frac{4}{4} - \frac{x}{4}\right)^2$$

$$= \left(1 - \frac{x}{4}\right)^2$$

$$= 1 - \frac{2x}{4} + \frac{x^2}{16}$$

$$= 1 - \frac{\overset{}{x}}{2} + \frac{x^2}{16} \quad Answer$$

A.9

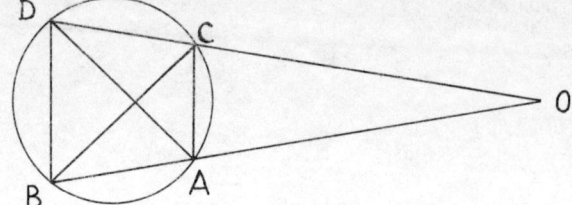

Fig. 3.24

Given: $AB = DB$

          $OD = OB$

To Prove: $OA = OC$

Construction: Join BC, AD

Proof: In $\triangle$'s BCD and ABD

          $D\widehat{A}B = B\widehat{C}D$ On same arc BD

          $C\widehat{D}B = A\widehat{B}D$ $\triangle ODB$ is isosceles

          $DB = AB$ Given: $\therefore$ $\triangle$'s are congruent A.A.S.

then $AB = CD$. Since $OB = OD$ and $OB - AB = OD - CD$

          $\therefore$ $OA = OC$ *Answer*

TEST PAPER 6

Q.1 Simplify

(a)
$$\frac{1}{\dfrac{1}{c^3} + \dfrac{1}{d^3} + \dfrac{1}{e^3}}$$

(b)
$$\frac{\dfrac{1}{c^4} + \dfrac{1}{c^5} + \dfrac{1}{c^6}}{c^4 + c^5 + c^6}$$

(c)
$$\frac{c^2 - d^2}{f}\left(1 - \frac{c + f}{c - d}\right)$$

Q.2 (a) Find the value by using common logarithms of the following:
$$(27 \cdot 03)^{-\frac{1}{2}} \times (0 \cdot 382)^{1 \cdot 27} \times 11 \cdot 15 \times (6 \cdot 85)^2$$

(b) Evaluate using Napierian logarithms:
$$(230 \cdot 6)^{-\frac{1}{2}} \times (0 \cdot 0573)^{0 \cdot 6}$$

Q.3 Find the value of x in:

$$Z = K + \frac{x(Z + K)}{Z + K + x}$$

Q.4 Solve the following equations:

(a)
$$2x^2 + \frac{5}{x} = 7$$

(b)
$$(3x + 1)(2x + 3) = 5x^2 - 9$$

Q.5 Draw the graph of $\frac{x^2}{16} + \frac{y^2}{9} = 1$ and using Simpson's Rules find the area under the curve to the x-axis, for values of $x = 3$ and 0.

Q.6 Find all the angles between $0°$ and $360°$ that will satisfy the following equation:

$$3 \cos 2\theta - 2 \cos \theta = 0$$

Q.7 Prove

(a)
$$\sin(A + B) + \sin(A - B) = 2 \sin A . \cos B$$

(b)
$$\frac{1 - \tan^2 A}{1 + \tan^2 A} = \frac{1}{\sec^2 A} - \frac{1}{\text{cosec}^2 A}$$

Q.8 A solid cone of 80cm base and 50cm slant height is melted down and remoulded into the form of a hollow sphere. The outside radius of the sphere is 50cm. Find the thickness of the material.

Q.9 $ABCD$ is a parallelogram and E is the mid-point of CD. The lines AE and BC are produced to meet at point F. Prove that:

(a) $AD = CF$
(b) Area of triangle $EFC = \frac{1}{4}$ Area of $ABCD$.

A.1 (a)
$$\frac{1}{\frac{1}{c^3} + \frac{1}{d^3} + \frac{1}{e^3}} = \frac{1}{\frac{d^3e^3 + c^3e^3 + c^3d^3}{c^3d^3e^3}}$$

$$= \frac{c^3d^3e^3}{d^3e^3 + c^3e^3 + c^3d^3} \quad Answer$$

(b)
$$\frac{\frac{1}{c^4} + \frac{1}{c^5} + \frac{1}{c^6}}{c^4 + c^5 + c^6} = \frac{\frac{c^2 + c + 1}{c^6}}{c^4(c^2 + c + 1)}$$

$$= \frac{\cancel{c^2 + c + 1}}{c^6 \times c^4 \cancel{(c^2 + c + 1)}}$$

$$= \frac{1}{c^{10}} \quad Answer$$

(c) $\dfrac{c^2 - d^2}{f} \cdot \left[1 - \dfrac{(c + f)}{(c - d)} \right] = \dfrac{(c + d)(c - d)}{f} \cdot \left[1 - \dfrac{(c + f)}{(c - d)} \right]$

$\qquad = \dfrac{(c + d)(c - d)}{f} - \dfrac{(c + d)\cancel{(c - d)}}{f} \cdot \dfrac{(c + f)}{\cancel{(c - d)}}$

$\qquad = \dfrac{(c + d)(c - d) - (c + d)(c + f)}{f}$

$\qquad = \dfrac{\cancel{c^2} + \cancel{cd} - \cancel{cd} - d^2 - \cancel{c^2} - cd - fd - fc}{f}$

$\qquad = \dfrac{-d^2 - cd - fd - fc}{f}$

$\qquad = \dfrac{-d(d + c) - f(d + c)}{f}$

$\qquad = \dfrac{-(d + f)(d + c)}{f}$ *Answer*

A.2 (a) $(27{\cdot}03)^{-\frac{1}{2}} \times (0{\cdot}382)^{1{\cdot}27} \times 11{\cdot}15 \times (6{\cdot}85)^2$

$\qquad \dfrac{(0{\cdot}382)^{1{\cdot}27} \times 11{\cdot}15 \times (6{\cdot}85)^2}{(27{\cdot}03)^{\frac{1}{2}}}$

$\qquad 0{\cdot}382^{1{\cdot}27} = 1{\cdot}27 \times \log 0{\cdot}382$
$\qquad\qquad\quad = 1{\cdot}27 \times \bar{1}{\cdot}5821$
$\qquad\qquad\quad = 1{\cdot}27 \times -0{\cdot}4179$

$\qquad\qquad\qquad\quad \begin{array}{r} -1 + 1 \\ \hline \end{array}$
$\qquad\qquad\quad = -0{\cdot}5307$
$\qquad\qquad\quad = \bar{1}{\cdot}4693$

$\qquad\qquad 6{\cdot}85^2 = 2 \times \log 6{\cdot}85$
$\qquad\qquad\quad\;\; = 2 \times 0{\cdot}8357$
$\qquad\qquad\quad\;\; = 1{\cdot}6714$

$\qquad\qquad 27{\cdot}03^{\frac{1}{2}} = \tfrac{1}{2} \times \log 27{\cdot}03$
$\qquad\qquad\qquad\;\; = \dfrac{1{\cdot}4319}{2}$
$\qquad\qquad\qquad\;\; = 0{\cdot}7159$

$\qquad \log 0{\cdot}382^{1{\cdot}27} = \bar{1}{\cdot}4693$
$\qquad\qquad \log 11{\cdot}15 = 1{\cdot}0472$
$\qquad\qquad \log 6{\cdot}85^2 = 1{\cdot}6714$
$\qquad\qquad\qquad\qquad\;\; \overline{2{\cdot}1879}$
$\qquad \log 27{\cdot}03^{\frac{1}{2}} = 0{\cdot}7159$
$\qquad\qquad\qquad\qquad\;\; \overline{1{\cdot}4720}$

Antilogging $= 29{\cdot}65$ *Answer (a)*

(b)
$$(230 \cdot 6)^{-\frac{1}{3}} \times (0 \cdot 0573)^{0 \cdot 6} = \frac{0 \cdot 0573^{0 \cdot 6}}{230 \cdot 6^{\frac{1}{3}}}$$

$$
\begin{aligned}
0 \cdot 0573^{0 \cdot 6} &= 0 \cdot 6 \times \log_e 0 \cdot 0573 \\
&= 0 \cdot 6 \times \bar{3} \cdot 1405 \\
&= 0 \cdot 6 \times -2 \cdot 8595 \\
&= \frac{-1 + 1}{-1 \cdot 7157} \\
&= \bar{2} \cdot 2843
\end{aligned}
$$

$$
\begin{aligned}
230 \cdot 6^{\frac{1}{3}} &= \tfrac{1}{3} \times \log_e 230 \cdot 6 \\
&= \tfrac{1}{3} \times 5 \cdot 4407 \\
&= \frac{5 \cdot 4407}{3} \\
&= 1 \cdot 8135
\end{aligned}
$$

$$
\begin{array}{ll}
\log_e 0 \cdot 0573^{0 \cdot 6} = & \bar{2} \cdot 2843 \\
\log_e 230 \cdot 6^{\frac{1}{3}} = & 1 \cdot 8135 \\
\log_e \text{ answer} & \overline{\bar{4} \cdot 4708} \\
\text{Antilogging} & \bar{5} \cdot 3948 \\
\text{Enter tables with} & \overline{1 \cdot 0760}
\end{array}
$$

$$= 0 \cdot 0293 \quad Answer \ (b)$$

A.3
$$Z = K + \frac{x(Z + K)}{Z + K + x}$$

$$Z - K = \frac{x(Z + K)}{Z + K + x}$$

$$(Z - K)(Z + K + x) = Zx + Kx$$

$$Z^2 + Zx - K^2 - Kx = Zx + Kx$$

$$Z^2 - K^2 = \cancel{Zx} - \cancel{Zx} + Kx + Kx$$

$$Z^2 - K^2 = 2Kx$$

$$\therefore \quad x = \frac{(Z + K)(Z - K)}{2K} \quad Answer$$

A.4 (a)
$$2x^2 + \frac{5}{x} = 7$$

$$2x^3 - 7x + 5 = 0$$

Factors of $5 = \pm 1 \pm 5$, substitute in above equation in place of x to obtain zero.

Try $x = 1$;

$$2 \times 1^3 - 7 \times 1 + 5 = 0$$

$\therefore \quad x = 1$ is a root and $(x - 1)$ is a factor

$$
\begin{array}{r}
x-1)\overline{2x^3 - 7x + 5}(2x^2 + 2x - 5 \\
2x^3 - 2x^2 \\
\hline
2x^2 - 7x + 5 \\
2x^2 - 2x \\
\hline
- 5x + 5 \\
- 5x + 5 \\
\hline
\end{array}
$$

$\left.\begin{array}{l} a = 2 \\ b = 2 \\ c = -5 \end{array}\right\} \qquad x = \dfrac{-2 \pm \sqrt{2^2 - (4 \times 2 \times -5)}}{2 \times 2}$

$$x = \frac{-2 \pm \sqrt{4 + 40}}{4}$$

$$x = \frac{-2 \pm \sqrt{44}}{4} = \frac{-2 \pm 6 \cdot 633}{4}$$

$$\therefore \quad x = \frac{-2 + 6 \cdot 633}{4} = \frac{4 \cdot 633}{4} = 1 \cdot 1582$$

OR

$$x = \frac{-2 - 6 \cdot 633}{4} = \frac{-8 \cdot 633}{4} = -2 \cdot 1582$$

$$x = 1, \ 1 \cdot 1582 \text{ OR } -2 \cdot 1582 \quad \textit{Answer}$$

(b)
$$(3x + 1)(2x + 3) = 5x^2 - 9$$
$$6x^2 + 11x + 3 = 5x^2 - 9$$
$$x^2 + 11x + 12 = 0$$

$\begin{array}{l} a = 1 \\ b = 11 \\ c = 12 \end{array} \qquad x = \dfrac{-11 \pm \sqrt{11^2 - (4 \times 1 \times 12)}}{2 \times 1}$

$$x = \frac{-11 \pm \sqrt{121 - 48}}{2}$$

$$x = \frac{-11 \pm \sqrt{73}}{2} = \frac{-11 \pm 8 \cdot 544}{2}$$

$$x = \frac{-11 - 8 \cdot 544}{2} = \frac{-19 \cdot 544}{2} = -9 \cdot 772$$

OR

$$x = \frac{-11 + 8 \cdot 544}{2} = \frac{-2 \cdot 456}{2} = -1 \cdot 228$$

$$\therefore \quad x = -9 \cdot 772 \text{ OR } -1 \cdot 228 \quad \textit{Answer}$$

A.5

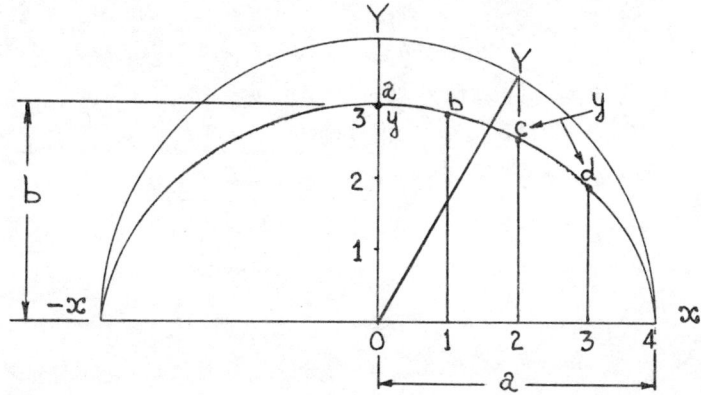

Fig. 3.25

$$\frac{x^2}{16} + \frac{y^2}{9} = 1$$

$$\frac{x}{4} + \frac{y}{3} = 1$$

$$\therefore \quad y = \frac{b}{a} \cdot Y = \frac{3}{4} \cdot Y$$

When

$$x = 0; \quad y = \tfrac{3}{4}\sqrt{(4^2 - 0^2)} = \tfrac{3}{4} \times 4 = 3$$

$$x = 1; \quad y = \tfrac{3}{4}\sqrt{(4^2 - 1^2)} = \tfrac{3}{4} \times 3{\cdot}873 = 2{\cdot}905$$

$$x = 2; \quad y = \tfrac{3}{4}\sqrt{(4^2 - 2^2)} = \tfrac{3}{4} \times 3{\cdot}464 = 2{\cdot}598$$

$$x = 3; \quad y = \tfrac{3}{4}\sqrt{(4^2 - 3^2)} = \tfrac{3}{4} \times 2{\cdot}646 = 1{\cdot}984$$

$$x = 4; \quad y = \tfrac{3}{4}\sqrt{(4^2 - 4^2)} = \tfrac{3}{4} \times 0 = 0$$

Using Simpson's 2nd Rule

$$\tfrac{3}{8} \cdot h(1 + 3 + 3 + 1) \quad h = 1$$

Ordinate	Length Ordinate	S.M's	Products
a	3	1	3·0
b	2·905	3	8·7150
c	2·598	3	7·7940
d	1·984	1	1·9840
			21·4930

$$\tfrac{3}{8}h(21{\cdot}493) = \tfrac{3}{8} \times 1 \times 21{\cdot}493 = \frac{64{\cdot}479}{8}$$

$$\therefore \quad \text{Area} = 8{\cdot}06 \text{ units}^2 \quad \textit{Answer}$$

A.6

$$3\cos 2\theta - 2\cos\theta = 0$$
$$3[2\cos^2\theta - 1] - 2\cos\theta = 0$$
$$6\cos^2\theta - 3 - 2\cos\theta = 0$$
$$6\cos^2\theta - 2\cos\theta - 3 = 0$$

$$\left.\begin{array}{l} a = 6 \\ b = -2 \\ c = -3 \end{array}\right\} \quad \theta = \frac{2 \pm \sqrt{4 - (4 \times 6 \times -3)}}{2 \times 6}$$

$$\theta = \frac{2 \pm \sqrt{4 + 72}}{12}$$

$$\theta = \frac{2 \pm \sqrt{76}}{12} = \frac{2 \pm 8\cdot7178}{12}$$

$$\theta = \frac{2 + 8\cdot7178}{12} = \frac{10\cdot7178}{12} = 0\cdot8931$$

OR

$$\theta = \frac{2 - 8\cdot7178}{12} = \frac{-6\cdot7178}{12} = -0\cdot5593$$

$$\therefore \quad \theta = 26°44', \text{ and } -56°$$

$$\therefore \quad \angle\theta = 26°44', 124°, 236°, 333°16' \quad \textit{Answer}$$

A.7 (a) $\sin(A + B) + \sin(A - B) = 2\sin A . \cos B$

LHS $= \sin A . \cos B + \cancel{\cos A . \sin B} + \sin A . \cos B - \cancel{\sin B . \cos A}$

$= 2\sin A . \cos B = $ RHS   *Answer (a)*

(b)

$$\frac{1 - \tan^2 A}{1 + \tan^2 A} = \frac{1}{\sec^2 A} - \frac{1}{\mathrm{cosec}^2 A}$$

$$\text{LHS} = \frac{1 - \dfrac{\sin^2 A}{\cos^2 A}}{1 + \dfrac{\sin^2 A}{\cos^2 A}}$$

$$= \frac{\dfrac{\cos^2 A - \sin^2 A}{\cancel{\cos^2 A}}}{\dfrac{\cos^2 A + \sin^2 A}{\cancel{\cos^2 A}}}$$

$$= \frac{\cos^2 A - \sin^2 A}{1} \quad (\text{because } \cos^2 A + \sin^2 A = 1)$$

$$= \frac{1}{\sec^2 A} - \frac{1}{\mathrm{cosec}^2 A} = \text{RHS} \quad \textit{Answer (b)}$$

A.8

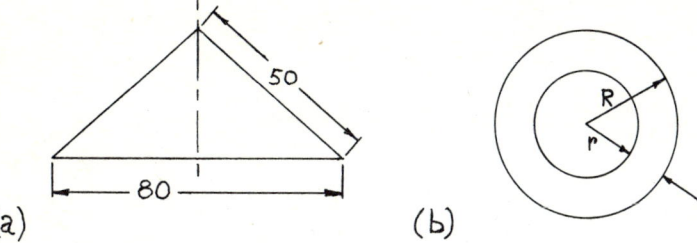

Fig. 3.26

Where:

$$R = 50cm$$

$$\text{Volume of cone} = \tfrac{1}{3}\pi r^2 h$$

$$\text{Volume of sphere} = \tfrac{4}{3}\pi r^3$$

Where $h = \sqrt{50^2 - 40^2} = \sqrt{(50 + 40)(50 - 40)}$

$$\therefore \quad h = \sqrt{90 \times 10} = \sqrt{900} = 30cm$$

Volume of cone = Volume large sphere − Volume small sphere

$$\therefore \quad \tfrac{1}{3}\pi r^2 h = (\tfrac{4}{3}\pi R^3 - \tfrac{4}{3}\pi r^3)$$

$$\tfrac{1}{3} \times \tfrac{22}{7} \times 40^2 \times 30 = \tfrac{4}{3}\pi(R^3 - r^3)$$

$$\tfrac{1}{3} \times \tfrac{22}{7} \times 40^2 \times 30 = \tfrac{4}{3} \times \tfrac{22}{7}(50^3 - r^3)$$

$$50^3 - r^3 = \tfrac{1}{\cancel{3}} \times \cancel{\tfrac{22}{7}} \times \overset{400}{\cancel{40^2}} \times 30 \times \tfrac{\cancel{3}}{4} \times \tfrac{7}{\cancel{22}}$$

$$50^3 - r^3 = 12\,000$$

$$\therefore \quad r^3 = 125\,000 - 12\,000$$

$$r = \sqrt[3]{113\,000} = 48\cdot34cm$$

$$\therefore \quad \text{Thickness of material} = 50 - 48\cdot34$$

$$= 1\cdot66cm \quad Answer$$

A.9

To Prove: (a) $AD = CF$
            (b) $Area \triangle EFC = \tfrac{1}{4}$ Area $ABCD$
Given: Above figure.
Construction: None.
Proof:

(a) In $\triangle$'s ADE and ECF

$$\angle ADE = \angle ECF \qquad \text{Alternate angles}$$

$$DE = EC \qquad \text{Given}$$

$$\angle AED = \angle CEF \qquad \text{Opposite angles}$$

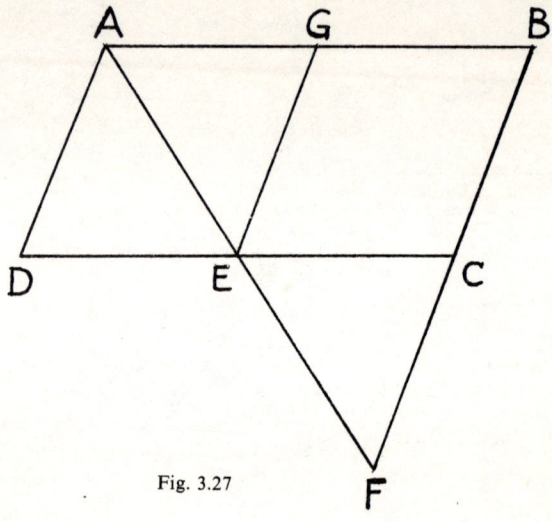

Fig. 3.27

$\therefore$ $\triangle$'s are congruent ASA

then $AD = CF$ *Answer* (a)

(b)   Construction: Draw $EG \parallel AD$ to meet AB at G.

Then

$$AG = GB$$

$$\triangle ADE = \triangle AGE \qquad \text{Construction}$$

$$\triangle ADE = \tfrac{1}{2}ADEG \qquad \text{Construction}$$

$$ADEG = \tfrac{1}{2}ABCD \qquad \text{Construction}$$

$$\therefore \quad \triangle ADE = \tfrac{1}{4}ABCD \qquad \text{Construction}$$

$$\triangle ADE = \triangle ECF \qquad \text{Congruent proved above.}$$

$$\therefore \quad \triangle EFC = \tfrac{1}{4}ABCD \qquad \textit{Answer (b)}$$

TEST PAPER 7

Q.1   Simplify, expressing the answer with positive indices without root signs.

(a)
$$\sqrt{\frac{(r^{\frac{4}{3}}s^5)}{(r^3 s^2)^2 s^4}}$$

(b)
$$\frac{\sqrt{\sqrt[3]{x^2 y^2}}}{\sqrt[4]{x^8 y^{2\cdot 5}}}$$

(c)
$$\left(\frac{27}{x^6}\right)^{-\frac{1}{3}}$$

Q.2   Solve using Napierian logarithms:

(a)
$$y \times 0.0562^{3.2} = 0.003821^{-2.1}$$

(b)
$$x = 0.678 \times \sqrt{\frac{0.00678}{6.78 \times 0.678}}$$

Q.3   Transpose for Z and evaluate:

$$Q = \frac{b}{2.87}\sqrt{\frac{f}{k}\left(R - \frac{Z}{S}\right)}$$

Where $R = 90\,000$, $Q = 55$, $b = 5.67$, $f = 32$, $k = 590$ and $S = 0.35$.

Q.4   The expression $x^3 + x^2 + ax + b$ is divisible by $(x + 7)$. When it is divided by $(x + 1)$ there is a remainder of 14. Find the values of a and b.

Q.5   Draw a line at 45° to a horizontal axis such that its ends are at 30mm and 60mm respectively from the axis. Calculate, using Simpson's Rules, the volume of solid evolved when this shape is rotated about the axis.

Q.6   A tower is viewed from a point A due east of the tower. The vertical angle at point A being 20°. At a second point B, 100m due south of A, the vertical angle was 18°; find the height of the tower.

Q.7   A conical float which operates a switch has a base diameter of 200mm and a height of 400mm, it floats freely with the apex 300mm below the surface of the fluid. Calculate the increased volume of fluid displaced by the float when, at the instant the switch operates, the apex goes to 350mm below the surface of the liquid.

Q.8   A sphere of 17mm radius has a cylindrical hole of 8mm radius cut out down the axis of the sphere. Calculate the volume of material remaining.

Q.9   A trapezium with sides of 10, 7, 15 and 8cm respectively, with the two longest sides parallel and the two short sides extended to form a triangle, is to be shown in diagrammatic form. Mathematically, find the area of the large triangle so formed.

A.1

$$\sqrt{\frac{(r^{\frac{8}{3}}s^5)}{(r^3s^2)^2 \cdot s^4}} = \left(\frac{(r^{\frac{8}{3}}s^5)}{r^6 s^4 \cdot s^4}\right)^{\frac{1}{2}}$$

$$= \frac{r^{\frac{4}{3}}s^{\frac{5}{2}}}{r^3 s^4}$$

$$= \frac{1}{r^{\frac{9-4}{3}}r^{\frac{8-5}{2}}}$$

$$= \frac{1}{r^{\frac{5}{3}}r^{\frac{3}{2}}} \quad Answer$$

(b)

$$\frac{\sqrt{\sqrt[3]{x^2y^2}}}{\sqrt[4]{x^8y^{2\cdot5}}} = \frac{(x^2y^2)^{\frac{1}{6}}}{(x^8y^{2\cdot5})^{\frac{1}{4}}}$$

$$= \frac{x^{\frac{1}{3}}y^{\frac{1}{3}}}{x^2y^{\frac{5}{8}}}$$

$$= \frac{1}{x^{\frac{6-1}{3}}y^{\frac{15-8}{24}}}$$

$$= \frac{1}{x^{\frac{5}{3}}y^{\frac{7}{24}}} \quad Answer$$

(c)

$$\left(\frac{27}{x^6}\right)^{-\frac{1}{3}} = \frac{1}{\left(\frac{27}{x^6}\right)^{\frac{1}{3}}} = \frac{1}{\frac{\sqrt[3]{27}}{x^2}}$$

$$= \frac{1}{\frac{3}{x^2}} = \frac{x^2}{3} \quad Answer$$

A.2 (a)

$$y \times 0{\cdot}0562^{3\cdot2} = 0{\cdot}003821^{-2\cdot1}$$

$$y = \frac{0{\cdot}003821^{-2\cdot1}}{0{\cdot}0562^{3\cdot2}}$$

$$y = \frac{1}{0{\cdot}0562^{3\cdot2} \times 0{\cdot}003821^{2\cdot1}}$$

$\log_e y = \log_e 1 \ - (3{\cdot}2\log_e 0{\cdot}0562 + 2{\cdot}1\log_e 0{\cdot}003821)$
$\quad\quad = 0{\cdot}0000 - (3{\cdot}2 \times \bar{3}{\cdot}1211 + 2{\cdot}1 \times \bar{6}{\cdot}4328)$
$\quad\quad = 0{\cdot}0000 - (3{\cdot}2 \times -2{\cdot}8789 + 2{\cdot}1 \times -5{\cdot}5672)$

$$\quad\quad\quad\quad\quad\quad\quad -1+1 \quad\quad -1+1$$
$\quad\quad = 0{\cdot}0000 - (\underline{-9}{\cdot}2125 + \underline{-}11{\cdot}6911)$
$\quad\quad = 0{\cdot}0000 - (\overline{10}{\cdot}7875 + \overline{12}{\cdot}3089)$
$\quad\quad = 0{\cdot}0000 - \overline{21}{\cdot}0964$
$\quad\quad = \overline{20}{\cdot}9036$

Antilogging

$$\begin{array}{r} \overline{20}{\cdot}9036 \\ \underline{20}{\cdot}7233 \\ \hline 0{\cdot}1803 \end{array}$$

$$\therefore \quad y = 1197 \times 10^6 \quad Answer$$

(b)
$$x = 0.678 \times \sqrt{\frac{0.00678}{6.78 \times 0.678}}$$

$$\log_e 0.00678 = \bar{3}.0062 \qquad \log_e 6.78 = 1.9140$$
$$\log_e 6.78 \times 0.678 = 1.5254 \qquad \log_e 0.678 = \bar{1}.6114$$
$$2\,\boxed{\bar{7}.4808} \qquad \log_e \text{b't'mline} = \overline{1.5254}$$
$$\overline{4.7404}$$
$$\log_e 0.678 = \bar{1}.6114$$
$$\log_e \text{Answer} = \overline{\bar{4}.3518}$$
$$\text{Antilogging} \qquad \overline{\bar{5}.3948}$$
$$\therefore \quad x = 0.9570 = 0.02604 \quad Answer$$

A.3
$$Q = \frac{b}{2.87}\sqrt{\frac{f}{k}\left(R - \frac{Z}{S}\right)}$$

$$\frac{2.87}{b}Q = \sqrt{\frac{f}{k}\left(R - \frac{Z}{S}\right)}$$

$$\left(\frac{2.87}{b}Q\right)^2 = \frac{f}{k}\left(R - \frac{Z}{S}\right)$$

$$\frac{k}{f}\left(\frac{2.87}{b}Q\right)^2 = \left(R - \frac{Z}{S}\right)$$

$$\frac{Z}{S} = R - \left(\frac{k}{f} \times \frac{2.87^2 Q^2}{b^2}\right)$$

$$Z = S\left[R - \left(\frac{k}{f} \times \frac{2.87^2 Q^2}{b^2}\right)\right]$$

Substituting values for $R = 90\,000$; $Q = 55$; $b = 5.67$; $f = 32$; $k = 590$ and $S = 0.35$.

$$Z = 0.35\left[90\,000 - \left(\frac{590}{32} \times \frac{2.87^2 \times 55^2}{5.67^2}\right)\right]$$

$$Z = 0.35\left[90\,000 - \left(\frac{590 \times 8.237 \times 3025}{32 \times 32.15}\right)\right]$$

$$Z = 0.35(90\,000 - 14\,290)$$

$$Z = 0.35 \times 75\,710$$

$$\therefore \quad Z = 26\,498.5 \quad Answer$$

A.4
$$x + 7)x^3 + x^2 + ax + b(x^2 - 6x + (42 + a)$$
$$\underline{x^3 + 7x^2}$$
$$-6x^2 + ax + b$$
$$\underline{-6x^2 - 42x}$$
$$(42 + a)x + b$$
$$\underline{(42 + a)x + 294 + 7a}$$
$$-294 - 7a + b = 0$$

$$\therefore \quad 7a - b = -294 \quad ①$$

93

$$x + 1)\overline{x^3 + x^2 + ax + b}(x^2 + a$$
$$\underline{x^3 + x^2}$$
$$+ ax + b$$
$$\underline{+ ax + a}$$
$$- a + b = 14$$

$$\therefore \quad a - b = -14 \quad ②$$

$$
\begin{array}{rl}
① & 7a - b = -294 \\
② & \underline{a - b = -\ 14} \\
& 6a \quad\quad = -280
\end{array}
$$

$$\therefore \quad a = \frac{-280}{6} = -46\tfrac{2}{3}$$

Substitute value of $a = -46\tfrac{2}{3}$ in ② above,

$$② \quad a - b = -14$$

$$-46\tfrac{2}{3} - b = -14$$

$$\therefore \quad b = -46\tfrac{2}{3} + 14 = -32\tfrac{2}{3}$$

$$Answer \quad a = -46\tfrac{2}{3}; \ b = -32\tfrac{2}{3}$$

A.5

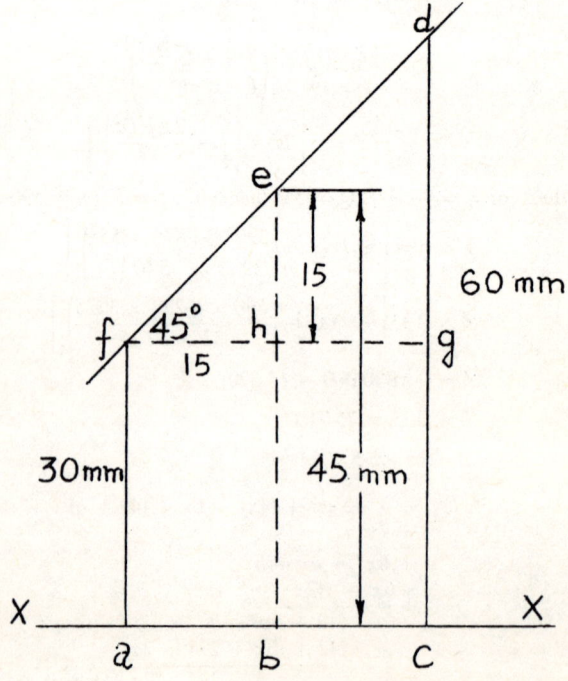

Fig. 3.28

$\triangle$'s *feh* and *fdg* are isosceles, therefore

$$fh = eh = 15\text{mm} \quad \text{and} \quad fg = dg = 30\text{mm} = fa$$

$\therefore$ *fg*, *ac* and *cg* = *fa* = 30mm

By the same reasoning *eh* + *bh* = 15 + 30 = *eb* = 45mm. From which, using Simpson's 1st Rule:

Ordinates	Radii	Area of circular section (πr^2)	(Simpson's Multipliers)			Products
a	30	$\pi \cdot 900$	×	1	=	$\pi \cdot 900$
b	45	$\pi \cdot 2025$	×	4	=	$\pi \cdot 8100$
c	60	$\pi \cdot 3600$	×	1	=	$\pi \cdot 3600$
						$\pi \cdot 12\,600$

$$\text{Volume} = \frac{h}{3}(a + 4b + c) \quad \text{where } h = \frac{30}{2} = 15\text{mm}$$

$$= \frac{15}{3} \times \pi \cdot \overset{4\,200}{\cancel{12\,600}}$$

$$= \overset{9}{\cancel{63}}\,000 \times \frac{22}{7}$$

$\therefore$ Volume = 198 000mm^3 or 0·000198m^3 *Answer*

A.6

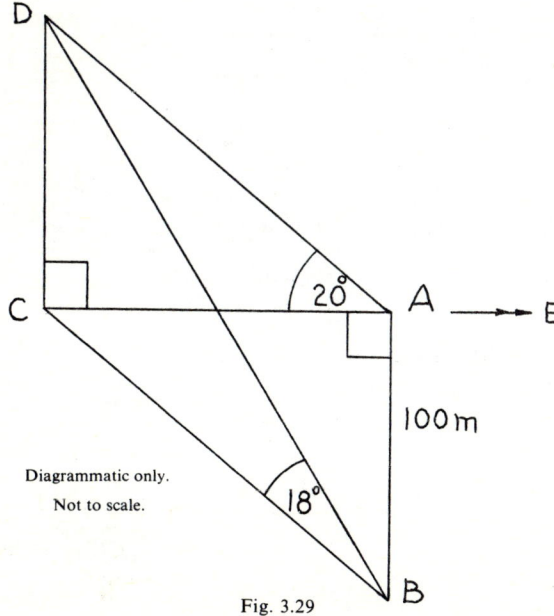

Diagrammatic only.
Not to scale.

Fig. 3.29

$$(BC)^2 - (AC)^2 = (AB)^2 = 100^2$$

But $BC = DC . \cot 18°$ and $AC = DC . \cot 20°$. Substituting these values in above, we get:

$$(DC . \cot 18°)^2 - (DC . \cot 20°)^2 = 100^2$$

$$(DC)^2(\cot^2 18° - \cot^2 20°) = 100^2$$

$$(a^2 - b^2) = (a + b)(a - b)$$

$$(DC)^2 = \frac{100^2}{(\cot^2 18° - \cot^2 20°)}$$

$$DC = \frac{100}{\sqrt{(\cot 18° + \cot 20°)(\cot 18° - \cot 20°)}}$$

$$DC = \frac{100}{\sqrt{(3·0777 + 2·7475)(3·0777 - 2·7475)}}$$

$$DC = \frac{100}{\sqrt{5·8252 \times 0·3302}} = \frac{100}{\sqrt{1·9414}} = \frac{100}{1·3933}$$

$$\therefore \quad DC = 71·76\text{m} \quad Answer$$

A.7

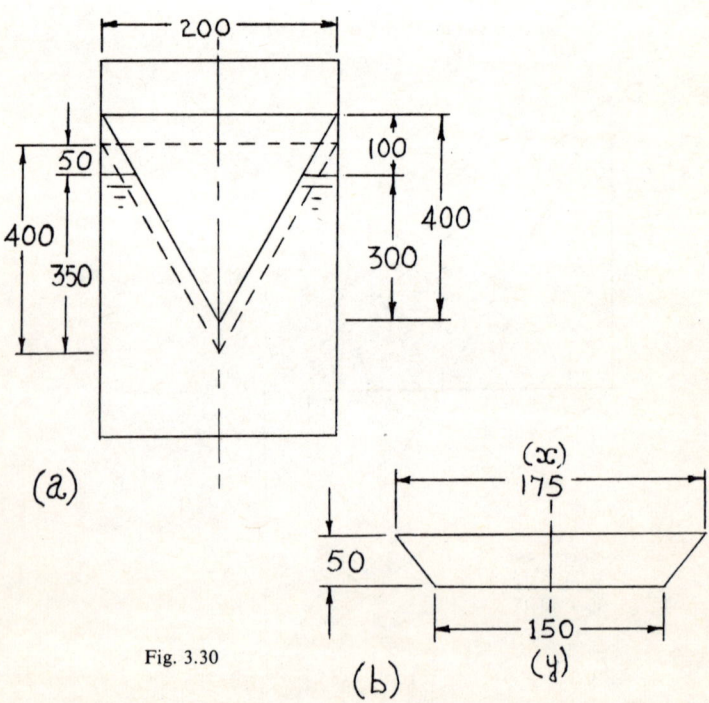

Fig. 3.30

By similar triangles:

$$\frac{\overset{2}{400}}{200} = \frac{300}{y} \qquad \therefore \quad y = \frac{300}{2} = 150mm$$

also

$$\frac{\overset{2}{400}}{200} = \frac{350}{x} \qquad \therefore \quad x = \frac{350}{2} = 175mm$$

Volume of frustrum $= \frac{1}{3}\pi h(R^2 + Rr + r^2)$

$= \frac{1}{3} \times 3 \cdot 1416 \times 50(87 \cdot 5^2 + 87 \cdot 5 \times 75 + 75^2)$

$= 52 \cdot 6(7656 + 6562 \cdot 5 + 5625)$

$= 52 \cdot 6 \times 19\ 843 \cdot 5$

$= 1\ 039\ 005 \cdot 66mm^3 = $ Volume of fluid displaced

$= \dfrac{1\ 039\ 005 \cdot 66}{10^9} = 0 \cdot 001\ 039m^3$ *Answer*

A.8

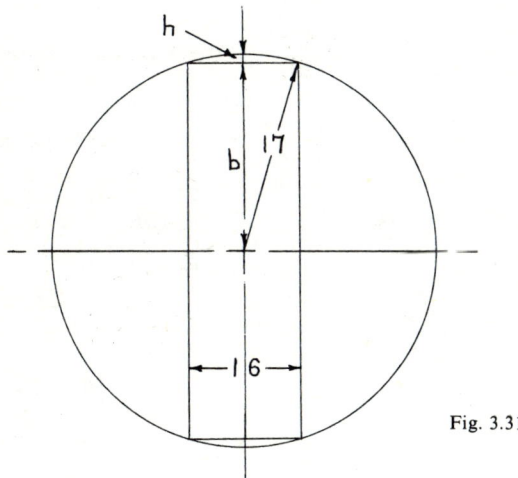

Fig. 3.31

Volume of sphere

$$= \tfrac{4}{3}\pi r^3$$

Volume of cylinder

$$= \pi r^2 h$$

Volume of segment

$$= \pi h^2(r - \tfrac{1}{3}h)$$

$h = 17 - b \quad$ where $b = \sqrt{17^2 - 8^2}$

$\therefore \quad b = \sqrt{289 - 64} = \sqrt{225} = 15mm$

and

$$h = 17 - 15 = 2mm$$

Volume of sphere remaining

$$= \text{Volume of sphere} - (\text{Volume of cylinder} + \text{Vol. 2-segments})$$

Or, using general formula;

$$\tfrac{4}{3}\pi(R^2 - r^2)^{\frac{3}{2}} = \tfrac{4}{3} \times \tfrac{22}{7} \times [(17 + 8)(17 - 8)]^{\frac{3}{2}}$$

$$= \frac{88}{21} \times (25 \times 9)^{\frac{3}{2}}$$

$$= \frac{88}{21} \times 225^{\frac{3}{2}}$$

$$= \frac{88}{21} \times 15^3$$

$$= \frac{88}{\cancel{21}_{7}} \times \cancel{3375}^{1125} = \frac{99\,000}{7} = 14\,142{\cdot}857\text{mm}^3$$

$$= \frac{14\,142{\cdot}857}{10^9} = 0{\cdot}00001414\text{m}^3 \quad \textit{Answer}$$

A.9

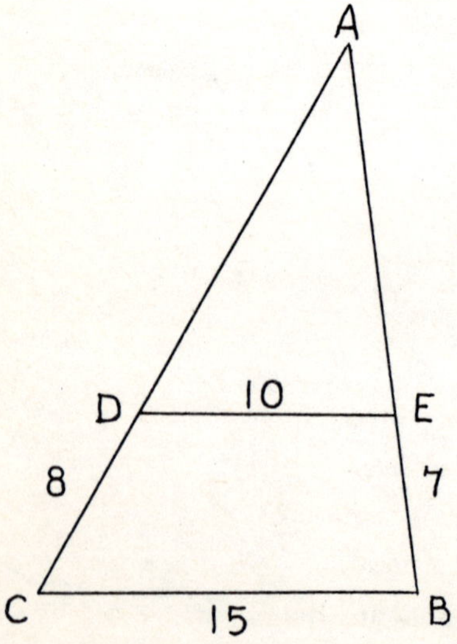

Fig. 3.32

Consider $\triangle$'s ADE and ABC

$\angle DAE = \angle CAB$ Common

$\angle ADE = \angle ACB$ Transversal

$\angle AED = \angle ABC$ Transversal

$\therefore$ $\triangle$'s ADE and ABC are similar, and the ratio of sides is constant.

$$\therefore \quad \frac{AE}{AB} = \frac{AD}{AC} = \frac{DE}{CB}$$

$$AE = AB \times \frac{DE}{CB} = (AE + 7) \times \frac{\cancel{10}^{2}}{\cancel{15}_{3}}$$

$$AE = \tfrac{2}{3}AE + \tfrac{14}{3}$$

$$3AE = 2AE + 14$$

$$3AE - 2AE = 14 \quad \therefore \quad AE = 14cm$$

$$AD = AC \times \frac{DE}{CB} = (AD + 8) \times \frac{\cancel{10}^{2}}{\cancel{15}_{3}}$$

$$AD = \tfrac{2}{3}AD + \tfrac{16}{3}$$

$$3AD = 2AD + 16$$

$$3AD - 2AD = 16 \qquad \therefore \quad AD = 16cm$$

To find area of triangle ABC

$$\left. \begin{array}{l} AB = 14 + 7 = 21cm \\ AC = 16 + 8 = 24cm \end{array} \right\}$$

and

also
$$CB = 15 \quad \text{given}$$

$$\text{Area of } \triangle ABC = \sqrt{s(s-a)(s-b)(s-c)}$$

where
$$s = \frac{a+b+c}{2} = \frac{15+24+21}{2}$$

$$\therefore \quad s = \tfrac{60}{2} = 30$$

$$\text{Area } \triangle ABC = \sqrt{30(30-15)(30-24)(30-21)}$$

$$= \sqrt{30 \times 15 \times 6 \times 9}$$

$$= \sqrt{24\,300}$$

$$\text{Area } \triangle ABC = 155.9cm^2 \quad Answer$$

TEST PAPER 8

Q.1 Simplify:

(a)
$$(m+n)^{\frac{1}{2}} \times \sqrt[5]{\frac{(m+n)^3}{(m^2-n^2)^2}}$$

(b)
$$\frac{7\tfrac{1}{3} - (3\tfrac{1}{2} \div 1\tfrac{8}{13}) + 4\tfrac{1}{2}}{6\tfrac{2}{7} - 2\tfrac{1}{2} + 3\tfrac{5}{14}}$$

Q.2 Evaluate for D in:

$$15(D + 0.05)^{1.35} = 144(0.05)^{1.45}$$

Q.3 Transpose for C in:

$$R = \frac{CWQ}{\sqrt{C^2 + W^2Q^2}}$$

Q.4 The sum of the sides of a rectangle is 25 units. Find the lengths of its sides, if the diagonal is 9.2 units.

Q.5 Three ships cross a channel 60 miles wide. Ships X and Y cross from A to B at 12 and 30 knots respectively. X leaves at 1100 hours and Y at 1200 hours. Z crosses from B to A and leaves at 1100 hours travelling at a speed of 24 knots. By means of a graph, find the time at which the three ships will meet. Also find their respective times of arrival.

Q.6 (a)   Prove with any acute angled triangle ABC that

$$\frac{a}{\sin A} = \frac{b}{\sin B} = \frac{c}{\sin C}$$

(b) X, Y and Z are three buoys each visible from the other two. From X; Y is $16°$ south of west and Z is $18°$ west of south. From Z; Y is $46°$ west of north. The distance YZ is 118 units. Calculate the distance X is from Y and Z.

Q.7   A trapezium is enclosed by a semi-circle of radius 2-units. The top parallel line is equal to the radius.
   (a)   Find the area of the trapezium.
   (b)   Using the Theorem of Pappus, calculate the volume of the frustrum of the cone formed when the figure is revolved about the vertical axis.

Q.8   An octagonal shaped figure has sides of one unit length. If 16% of the area of the octagon was to be cut off, how far from the centre would the line be if it was cut parallel to one side.

Q.9   In Fig. 3.33, if AD is parallel to BC; what is angle θ?

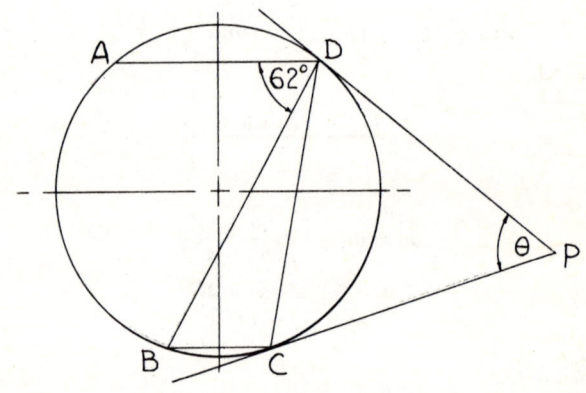

Fig. 3.33

A.1 (a)

$$(m + n)^{\frac{1}{5}} \times \sqrt[5]{\frac{(m + n)^3}{(m^2 - n^2)^2}} = (m + n)^{\frac{1}{5}} \times \sqrt[5]{\frac{(m + n)(m + n)(m + n)}{(m + n)(m - n)(m + n)(m - n)}}$$

$$= (m + n)^{\frac{1}{5}} \times \sqrt[5]{\frac{(m + n)}{(m - n)(m - n)}}$$

$$= \frac{(m + n)^{\frac{1}{5}} \times (m + n)^{\frac{1}{5}}}{(m - n)^{\frac{2}{5}}}$$

$$= \frac{(m + n)^{\frac{2}{5}}}{(m - n)^{\frac{2}{5}}} \quad \textit{Answer}$$

(b)

$$\frac{7\frac{1}{3} - (3\frac{1}{2} \div 1\frac{8}{13}) + 4\frac{1}{2}}{6\frac{2}{7} - 2\frac{1}{2} + 3\frac{5}{14}} = \frac{\frac{22}{3} - (\frac{7}{2} \times \frac{13}{21}) + \frac{9}{2}}{\frac{88}{14} - \frac{35}{14} + \frac{47}{14}}$$

$$= \frac{\frac{22}{3} - \frac{91}{42} + \frac{9}{2}}{\frac{100}{14}}$$

$$= \frac{\frac{308 - 91 + 189}{42}}{\frac{100}{14}} = \frac{\frac{406}{42}}{\frac{100}{14}}$$

$$= \frac{\overset{203}{\cancel{406}}}{\underset{3}{\cancel{42}}} \times \frac{\overset{1}{\cancel{14}}}{\underset{50}{\cancel{100}}} = \frac{203}{150}$$

$$= 1\frac{53}{150} \quad Answer$$

A.2

$$15(D + 0\cdot05)^{1\cdot35} = 144(0\cdot05)^{1\cdot45}$$

$$\log 15 + 1\cdot35 \times \log(D + 0\cdot05) = \log 144 + 1\cdot45 \times \log 0\cdot05$$

$$1\cdot1761 + 1\cdot35 \log(D + 0\cdot05) = 2\cdot1584 + 1\cdot45 \times \bar{2}\cdot6990$$

$$1\cdot35 \log(D + 0\cdot05) = 2\cdot1584 + \bar{2}\cdot1135 - 1\cdot1761$$

$$1\cdot35 \log(D + 0\cdot05) = 0\cdot2719 - 1\cdot1761$$

$$\left. \log(D + 0\cdot05) = \frac{\bar{1}\cdot0958}{1\cdot35} = \frac{-0\cdot9042}{1\cdot35} = \begin{matrix} -1 +1 \\ -0\cdot6697 \\ = \bar{1}\cdot3303 \end{matrix} \right)$$

$$\log(D + 0\cdot05) = \bar{1}\cdot3303$$

Antilogging $\quad D + 0\cdot05 = 0\cdot2139$

$$\therefore \quad D = 0\cdot2139 - 0\cdot05$$

$$D = 0\cdot1639 \quad Answer$$

A.3

$$R = \frac{CWQ}{\sqrt{C^2 + W^2Q^2}}$$

Squaring both sides

$$R^2 = \frac{C^2W^2Q^2}{C^2 + W^2Q^2}$$

$$R^2C^2 + R^2W^2Q^2 = C^2W^2Q^2$$

$$R^2W^2Q^2 = C^2W^2Q^2 - R^2C^2$$

$$R^2W^2Q^2 = C^2(W^2Q^2 - R^2)$$

$$C^2 = \frac{R^2W^2Q^2}{W^2Q^2 - R^2}$$

$$C = \sqrt{\frac{R^2W^2Q^2}{W^2Q^2 - R^2}}$$

$$C = \frac{RWQ}{\sqrt{W^2Q^2 - R^2}}$$

$$\therefore \quad C = \frac{RWQ}{\sqrt{(WQ + R)(WQ - R)}} \quad \textit{Answer}$$

A.4

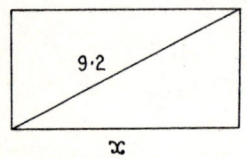

$$2x + 2y = 25 \quad \text{①}$$
$$x^2 + y^2 = 9{\cdot}2^2 \quad \text{②}$$
$$= 84{\cdot}64$$

Fig. 3.34

From ① $x + y = 12{\cdot}5$ $\quad \therefore x = 12{\cdot}5 - y$. Substituting in ② for x, we get:

$$(12{\cdot}5 - y)^2 + y^2 = 9{\cdot}2^2$$
$$12{\cdot}5^2 - 25y + y^2 + y^2 = 84{\cdot}64$$
$$2y^2 - 25y + 156{\cdot}30 = 84{\cdot}64$$
$$2y^2 - 25y + 71{\cdot}66 = 0$$

$$\left.\begin{array}{l} a = 2 \\ b = -25 \\ c = 71{\cdot}66 \end{array}\right\} \quad y = \frac{25 \pm \sqrt{-25^2 - (4 \times 2 \times 71{\cdot}66)}}{2 \times 2}$$

$$y = \frac{25 \pm \sqrt{625 - 573{\cdot}28}}{4}$$

$$y = \frac{25 \pm \sqrt{51{\cdot}72}}{4} = \frac{25 \pm 7{\cdot}191}{4}$$

$$\therefore \quad y = \frac{32{\cdot}191}{4} = 8{\cdot}047 \quad \left(\begin{array}{c} \text{As we have made } y \text{ the shortest side, this must be the value} \\ \text{for } x. \end{array}\right)$$

OR $\quad y = \dfrac{17{\cdot}809}{4} = 4{\cdot}453$

$$\textit{Answer:} \quad \left.\begin{array}{l} x = 8{\cdot}047 \text{ units} \\ y = 4{\cdot}453 \text{ units} \end{array}\right\}$$

A.5

		Leaves	Arrives
$X = \dfrac{60}{12} = 5$ hours		1100 hours	1600 hours
$Y = \dfrac{60}{30} = 2$ hours		1200	1400
$Z = \dfrac{60}{24} = 2\frac{1}{2}$ hours		1100	1330

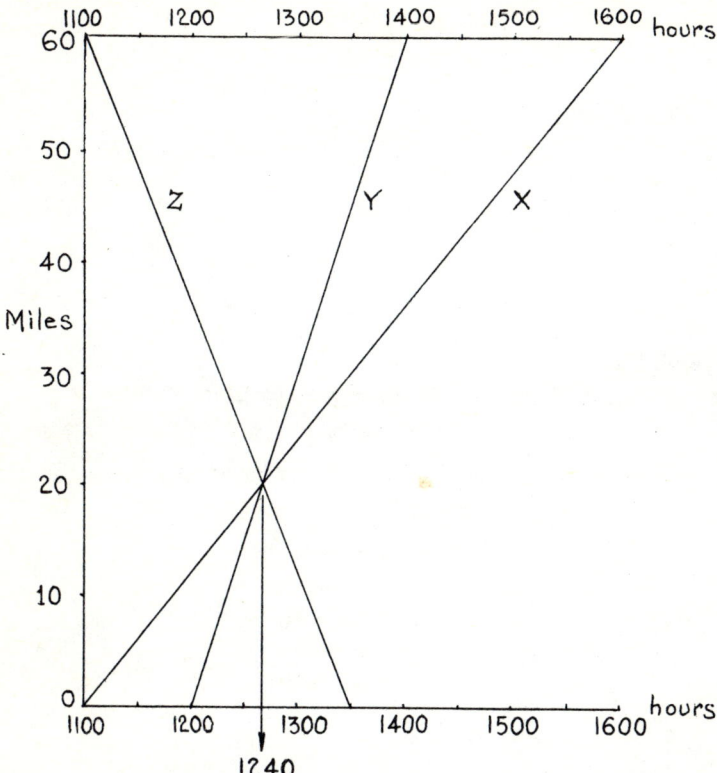

Fig. 3.35

Answer: Vessels will meet at 1240 hours

X arrives 1600 hours ⎫
Y arrives 1400 hours ⎬
Z arrives 1330 hours ⎭

A.6

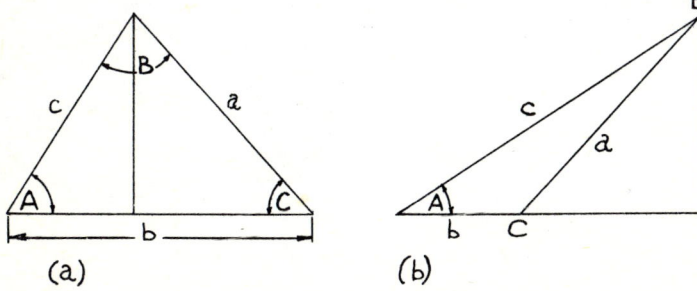

Fig. 3.36

$$\sin A = \frac{y}{c} \quad \therefore \quad y = c \cdot \sin A \quad \text{(Fig. 3.36(a))}$$

$$\sin C = \frac{y}{a} \quad \therefore \quad y = a \cdot \sin C$$

From which

$$a \cdot \sin C = c \cdot \sin A$$

Transposing

$$\frac{a}{\sin A} = \frac{c}{\sin C}$$

Further, using Fig. 3·36(b)

$$\angle ACB + \angle BCD = 180° = \text{Supplementary Angles.}$$

$$\therefore \quad \angle ACB = \angle BCD = \sin C$$

Because the sine is positive in both the first and second quadrants of the circle diagram.

$$\therefore \quad \frac{a}{\sin A} = \frac{b}{\sin B} = \frac{c}{\sin C}$$

The Sine Rule

(b)

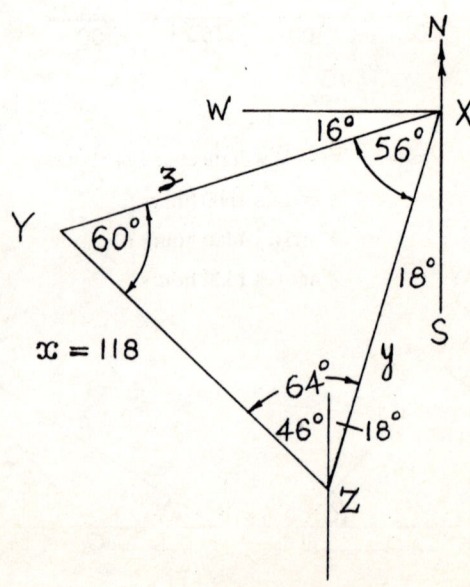

Fig. 3.37

Using the Sine Rule:

$$\frac{x}{\sin X} = \frac{y}{\sin Y} \quad \therefore \quad \frac{118}{\sin 56°} = \frac{y}{\sin 60°}$$

and

$$y = \frac{118 \cdot \sin 60°}{\sin 56°}$$

$$= \frac{118 \times 0\cdot866}{0\cdot829}$$

$$y = \frac{102\cdot188}{0\cdot829} = 123\cdot2 \ units$$

also

$$\frac{z}{\sin Z} = \frac{x}{\sin X} \quad \therefore \quad z = \frac{x \cdot \sin Z}{\sin X}$$

$$z = \frac{118 \cdot \sin 64°}{\sin 56°}$$

$$z = \frac{118 \times 0\cdot8988}{0\cdot829}$$

$$z = \frac{106\cdot058}{0\cdot829} = 127\cdot93 \ units$$

Answer (a) See Proof.

(b) X is 127·93 units from Y; ⎫
and 123·2 units from Z. ⎬

A.7 (a)

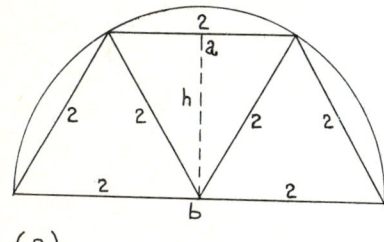

(a)

Fig. 3.38

To find area of trapezium

$$\text{Area} = \left(\frac{a + b}{2}\right) \cdot h$$

Where

$$h = \sqrt{2^2 - 1^2}$$

$$= \sqrt{3} = 1\cdot732 \ units$$

$$\therefore \quad Area = \left(\frac{2 + 4}{2}\right) \times 1\cdot732$$

$$= 3 \times 1\cdot732 = 5\cdot196 \ units^2$$

105

(b) To find volume of frustrum, using 2nd Theorem of Pappus.

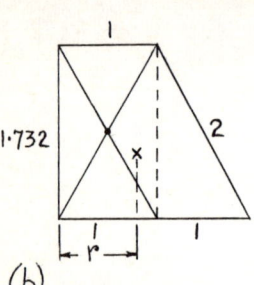

(b)

Volume = Area × Distance Travelled

by C.G (centroid)

= Area × $2\pi r$

Where $r = \dfrac{\text{Sum of moments}}{\text{Sum of areas}}$

$$r = \frac{(1\cdot732 \times 1 \times 0\cdot5) + (0\cdot866 \times 1\cdot33)}{1\cdot732 + 0\cdot866}$$

$$r = \frac{0\cdot866 + 1\cdot154}{2\cdot598} = \frac{2\cdot020}{2\cdot598} = 0\cdot78 \text{ units}$$

$$\therefore \quad Volume = \frac{5\cdot196}{2} \times 2 \times \pi \times r$$

$$= 2\cdot598 \times 6\cdot2832 \times 0\cdot78$$

$$= 12\cdot7008 \text{ units}^3$$

Answer (a) Area = 5·196 units² $\Big\}$
(b) Volume = 12·70 units³

A.8

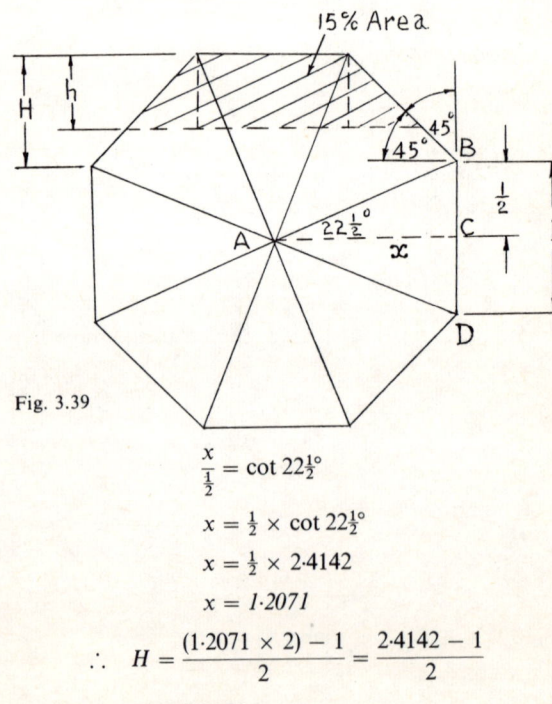

Fig. 3.39

$$\frac{x}{\frac{1}{2}} = \cot 22\tfrac{1}{2}°$$

$$x = \tfrac{1}{2} \times \cot 22\tfrac{1}{2}°$$

$$x = \tfrac{1}{2} \times 2\cdot4142$$

$$x = 1\cdot2071$$

$$\therefore \quad H = \frac{(1\cdot2071 \times 2) - 1}{2} = \frac{2\cdot4142 - 1}{2}$$

$$H = 0\cdot7071$$

$$\text{Area of } \triangle ABD = \frac{1 \cdot 2071 \times 1}{2} = 0 \cdot 6035 \text{ units}^2$$

$$\text{Area of octagon} = 0 \cdot 6035 \times 8 = 4 \cdot 8280 \text{ units}^2$$

$$15\% \text{ of area of octagon} = \frac{4 \cdot 828 \times 15}{100} = 0 \cdot 7242 \text{ units}^2$$

Area of small trapezium = 15% of area of octagon

$$\therefore \quad h^2 + (h \times 1) = 0 \cdot 7242$$

Transposing $\quad h^2 + h - 0 \cdot 7242 = 0 \quad$ [a quadratic equation]

$$\left. \begin{array}{l} a = 1 \\ b = 1 \\ c = -0 \cdot 7242 \end{array} \right\} \quad h = \frac{-1 \pm \sqrt{1^2 - (4 \times 1 \times -0 \cdot 7242)}}{2 \times 1}$$

$$h = \frac{-1 \pm \sqrt{3 \cdot 8968}}{2} = \frac{-1 \pm 1 \cdot 974}{2}$$

$$h = \frac{-1 - 1 \cdot 974}{2} = \cancel{-1 \cdot 487}$$

OR

$$h = \frac{0 \cdot 974}{2} = 0 \cdot 487$$

$$\therefore \quad \text{Distance of line from centre} = 0 \cdot 7071 - 0 \cdot 487 + 0 \cdot 5$$

$$= 0 \cdot 7201 \text{ units} \quad \textit{Answer}$$

A.9 AD is parallel to BC

$$\therefore \angle DBC = 62^{\circ} \quad \text{Alternate angles.}$$

$$\angle DBC = \angle PDC$$

$\therefore \quad \angle PDC = 62^{\circ} \quad \Big\{$ The angle which the chord makes with the tangent is equal

$\angle PDC = \angle PCD \quad \Big\{$ to the angle in the alternate segment

$\therefore \quad \triangle PDC$ is isosceles $\Big\{$ From a point outside a circle only two equal tangents can be drawn

$$\therefore \quad \angle PCD = 62^{\circ}$$

$$\angle \theta = 180^{\circ} - (62^{\circ} + 62^{\circ})$$

$$\therefore \quad \angle \theta = 56^{\circ} \quad \textit{Answer}$$

TEST PAPER 9

Q.1 Find the value of 'a' in the following:

(a) $8^{0 \cdot 5a} = 40$;   (b) $e^{4a} = 17 \cdot 8^{2 \cdot 6}$;   (c) $a^{3 \cdot 5} = 4a^2$;

(d) $\sqrt[5]{a} = e^{5 \cdot 28}$.

Q.2 Evaluate using common logarithms:

(a) $14.96^{\frac{2}{3}} \times 0.555^{-\frac{1}{2}} \times 0.015^{0.5}$

(b) Evaluate using Napierian logarithms:

$$93^2(\log_e 1.45 + \log_e 2.23)$$

Q.3 Transpose for h in:

$$P = A^3 . \sqrt{\dfrac{W^2}{\dfrac{d}{h} + \dfrac{h}{t}}}$$

Q.4 Draw three straight line graphs of values:

(1) $y = 2x - 10$ (2) $y = \frac{2}{5}x - 4$ (3) $y = -\frac{7}{9}x - 8$

and then find the area of the triangle so formed.

Q.5 A barrel has the following dimensions: Length of barrel $8\frac{5}{6}$ units; End diameters $6\frac{5}{6}$ units; $\frac{1}{4}$-length diameters $8\frac{1}{6}$ units; Mid-length diameter $8\frac{5}{6}$. Using Simpson's Rule, calculate the volume of the barrel.

Q.6 An observer on a cliff 30m high sights a vessel bearing due north at an angle of depression of 15°. After steering due east the bearing changes to N40°E. Find (a) The distance travelled between the first and second bearings. (b) The new angle of depression.

Q.7 If $\tan \theta = 3$, prove that $2 \sin^2 \theta + 3 \cos^2 \theta = 2\frac{1}{10}$.

Q.8 A pentagon whose distance from the centre to one corner is 3-units, has a slice cut off at an angle of 30° to the side from one corner. Find the ratio between the area of the pentagon and the portion sliced off.

Q.9 A wire loop of 14mm radius, had a straight piece of wire attached to the loop in the form of a chord (AB) to the circle and 26mm in length. Calculate (a) The distance of the wire from the centre 'O'; (b) The angle AOB; (c) The length of the large arc AEB.

A.1   (a)

$$8^{0.5a} = 40$$

$$0.5a \times \log 8 = \log 40$$

$$0.5a = \frac{\log 40}{\log 8}$$

$$0.5a = \frac{1.6021}{0.9031} = 1.774$$

$$a = \frac{1.774}{0.5} = 3.548$$

Answer $a = 3.548$

(b)
$$e^{4a} = 17 \cdot 8^{2 \cdot 6}$$
$$4a \times \log 2 \cdot 7183 = 2 \cdot 6 \log 17 \cdot 8$$
$$4a \times 0 \cdot 4343 = 2 \cdot 6 \times 1 \cdot 2504$$
$$4a = \frac{3 \cdot 2510}{0 \cdot 4343} = 7 \cdot 48$$
$$a = \frac{7 \cdot 48}{4} = 1 \cdot 87$$
Answer $a = 1 \cdot 87$

(c)
$$a^{3 \cdot 5} = 4a^2$$
$$3 \cdot 5 \log a = 2 \log a + \log 4$$
$$3 \cdot 5 \times \log a = 2 \log a + 0 \cdot 6021$$
$$3 \cdot 5 \log a - 2 \log a = 0 \cdot 6021$$
$$1 \cdot 5 \log a = 0 \cdot 6021$$
$$\log a = \frac{0 \cdot 6021}{1 \cdot 5} = 0 \cdot 4014$$
Antilogging $a = 2 \cdot 52$ *Answer*

(d)
$$\sqrt[5]{a} = e^{5 \cdot 3}$$
$$\tfrac{1}{5} \log a = 5 \cdot 3 \log 2 \cdot 7183$$
$$\tfrac{1}{5} \log a = 5 \cdot 3 \times 0 \cdot 4343$$
$$\tfrac{1}{5} \log a = 2 \cdot 3018$$
$$\log a = 5 \times 2 \cdot 3018$$
$$\log a = 11 \cdot 5090$$
Antilogging $a = 32 \cdot 28 \times 10^{10}$ *Answer*

A.2 (a)
$$14 \cdot 96^{\frac{2}{3}} \times 0 \cdot 555^{-\frac{1}{2}} \times 0 \cdot 015^{0 \cdot 5}$$

$\log 14 \cdot 96 =$	$1 \cdot 1750$	$\log 0 \cdot 555^{-\frac{1}{2}} = -\tfrac{1}{2} \times \bar{1} \cdot 7443$
	2	$= -\tfrac{1}{2} \times -0 \cdot 2557$
3	$\overline{2 \cdot 3500}$	$= \dfrac{-0 \cdot 2557}{-2}$

(i) $\log 14 \cdot 96^{\frac{2}{3}} = 0 \cdot 7833$

(ii) $\log 0 \cdot 555^{-\frac{2}{3}} = 0 \cdot 1278$ $= +0 \cdot 1278$ (ii)

(iii) $\log 0 \cdot 015^{0 \cdot 5} = \bar{1} \cdot 0880$

$\overline{\bar{1} \cdot 9991}$ $\log 0 \cdot 015^{0 \cdot 5}$

Antilogging $= \dfrac{\bar{2} \cdot 1761}{2}$ $[0 \cdot 5 = \tfrac{1}{2}]$

Answer $= 0 \cdot 9979$ $= \bar{1} \cdot 0880$ (iii)

(b)

$$93^2(\log_e 1.45 + \log_e 2.23) = 93^2(0.3716 + 0.8020)$$
$$= 93^2 \times 1.1736$$
$$= 2 \times \log_e 93 + \log_e 1.1736$$
$$= 9.0652 + 0.1601$$
$$= 9.2253$$

$$\log_e 9.3 = 2.2300$$
$$\log_e 10^1 = 2.3026$$
$$\overline{4.5326}$$
$$2$$
$$\log_e 93^2 = \overline{9.0652}$$
$$\log_e 1.1736 = 0.1601$$
$$\log_e \text{ answer} = \overline{9.2253}$$
$$\text{Antilogging} \quad 9.2103$$
$$\overline{0.0150}$$

$$Answer = 10\,150$$

A.3

$$P = A^3 \cdot \sqrt{\dfrac{W^2}{\dfrac{d}{h} + \dfrac{h}{t}}}$$

$$P^2 = A^6 \cdot \left(\dfrac{W^2}{\dfrac{d}{h} + \dfrac{h}{t}}\right)$$

$$\dfrac{d}{h} + \dfrac{h}{t} = \dfrac{A^6 \cdot W^2}{P^2}$$

$$\dfrac{d}{h} = \dfrac{A^6 \cdot W^2}{P^2} - \dfrac{h}{t}$$

$$d = h \cdot \left(\dfrac{A^6 \cdot W^2}{P^2} - \dfrac{h}{t}\right)$$

$$d = \dfrac{h \cdot A^6 W^2}{P^2} - \dfrac{h^2}{t}$$

$$\dfrac{h^2}{t} - \dfrac{h \cdot A^6 W^2}{P^2} + d = 0$$

$$\left.\begin{array}{l} a = \dfrac{1}{t} \\[2mm] b = \dfrac{-A^6 W^2}{P^2} \\[2mm] c = d \end{array}\right\} \quad h = \dfrac{\dfrac{A^6 W^2}{P^2} \pm \sqrt{-\left(\dfrac{A^6 W^2}{P^2}\right)^2 - \left(4 \times \dfrac{1}{t} \times d\right)}}{2 \times \dfrac{1}{t}}$$

$$h = \frac{\dfrac{A^6 W^2}{P^2} \pm \sqrt{\dfrac{A^{12} W^4}{P^4} - \dfrac{4}{t}d}}{\dfrac{2}{t}}$$

$$h = \frac{t}{2}\left(\frac{A^6 W^2}{P^2} \pm \sqrt{\frac{A^{12} W^4}{P^4} - \frac{4}{t}d}\right) \quad Answer$$

A.4

① $y = 2x - 10$

x	0	2	4
$2x$	0	4	8
-10	-10	-10	-10
y	-10	-6	-2

② $y = \frac{2}{5}x - 4$

x	0	2	4
$\frac{2}{5}x$	0	$\frac{4}{5}$	$\frac{8}{5}$
-4	-4	-4	-4
y	-4	$-3\frac{1}{5}$	$-2\frac{2}{5}$

③ $y = -\frac{7}{9}x - 8$

x	0	2	4
$-\frac{7}{9}x$	0	$-1\frac{5}{9}$	$-3\frac{1}{9}$
-8	-8	-8	-8
y	-8	$-9\frac{5}{9}$	$-11\frac{1}{9}$

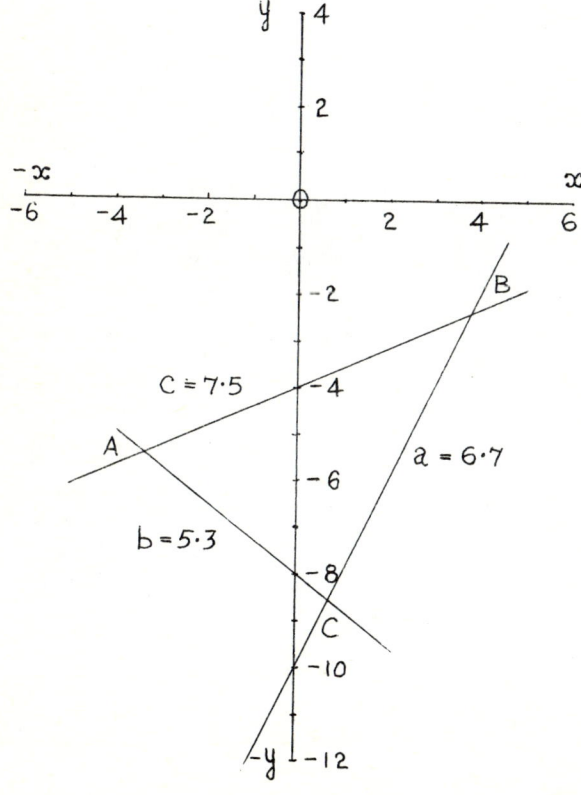

Fig. 3.40

Area $\triangle ABC$

$= \sqrt{s(s-a)(s-b)(s-c)}$

$= \sqrt{9{\cdot}75(9{\cdot}75 - 6{\cdot}7)(9{\cdot}75 - 5{\cdot}3)(9{\cdot}75 - 7{\cdot}5)}$

$= \sqrt{9{\cdot}75 \times 3{\cdot}05 \times 4{\cdot}45 \times 2{\cdot}25}$

$= \sqrt{298} = 17{\cdot}26 \, \text{units}^2$ *Answer*

Where

$s = \dfrac{a+b+c}{2}$

$s = \dfrac{6{\cdot}7 + 5{\cdot}3 + 7{\cdot}5}{2}$

$\therefore \quad s = \dfrac{19{\cdot}5}{2} = 9{\cdot}75$

A.5 Using Simpson's 1st Rule: $\dfrac{h}{3}[1 + 4 + 2 + 4 + 1]$

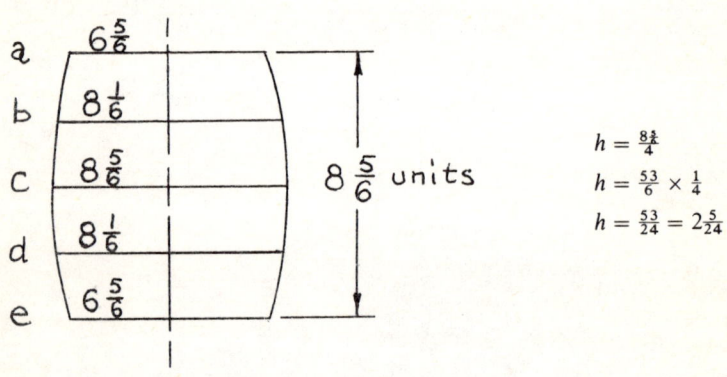

$h = \frac{8\frac{2}{3}}{4}$

$h = \frac{53}{6} \times \frac{1}{4}$

$h = \frac{53}{24} = 2\frac{5}{24}$

Fig. 3.41

Ordinate	Radii	πr^2	Areas	S.M's	Products
a	$\frac{41}{12}$	$\frac{22}{7} \times (\frac{41}{12})^2 = 36{\cdot}68$		1	36·68
b	$\frac{49}{12}$	$\frac{22}{7} \times (\frac{49}{12})^2 = 52{\cdot}40$		4	209·60
c	$\frac{53}{12}$	$\frac{22}{7} \times (\frac{53}{12})^2 = 61{\cdot}31$		2	122·62
d	$\frac{49}{12}$	$\frac{22}{7} \times (\frac{49}{12})^2 = 52{\cdot}40$		4	209·60
e	$\frac{41}{12}$	$\frac{22}{7} \times (\frac{41}{12})^2 = 36{\cdot}68$		1	36·68
					615·18

$\text{Volume} = \dfrac{h}{3} \times 615{\cdot}18$

$= \dfrac{53}{24} \times \dfrac{1}{3} \times \dfrac{615{\cdot}18}{1}$

$= \dfrac{32\,604{\cdot}54}{72}$

$\therefore \quad \text{Volume} = 452{\cdot}84 \, \text{units}^3$ *Answer*

A.6

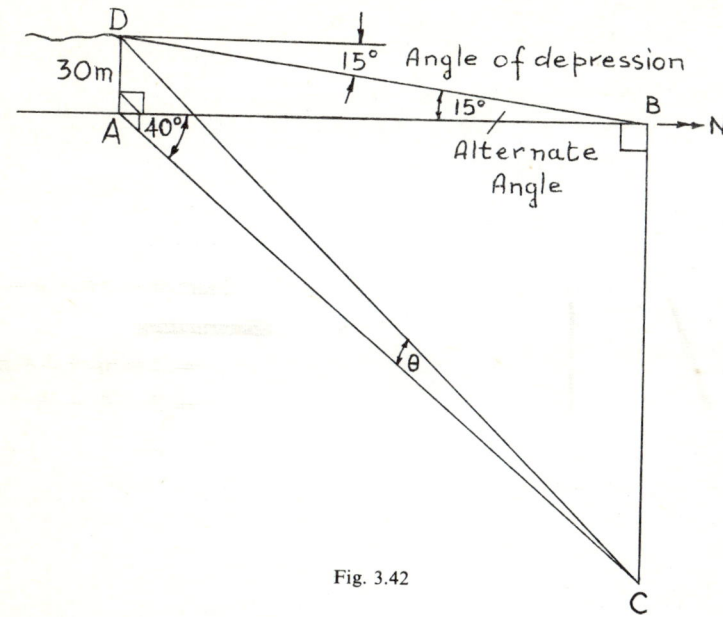

Fig. 3.42

(a) To find AB

$$\frac{AB}{AD} = \frac{AB}{30} = \cot 15°$$

$$\therefore \quad AB = 30 \times \cot 15°$$

$$AB = 30 \times 3 \cdot 7321 = 111 \cdot 963m$$

To find BC (Distance travelled)

$$\frac{BC}{AB} = \frac{BC}{111 \cdot 963} = \tan 40°$$

$$\therefore \quad BC = 111 \cdot 963 \times \tan 40°$$

$$BC = 111 \cdot 963 \times 0 \cdot 8391 = 93 \cdot 948m$$

(b) To find $\tan \theta$ (New Angle of Depression)

$$\tan \theta = \frac{AD}{AC} \quad \text{where } AC = BC \times \text{cosec } 40°$$

$$= 93 \cdot 948 \times 1 \cdot 5557$$

$$= 146 \cdot 155m$$

$$\therefore \quad \tan \theta = \frac{30}{146 \cdot 155} = 0 \cdot 2052 = 11°\,36'$$

Answer (a) 93·948m ⎫
⎬
(b) 11°36' ⎭

A.7

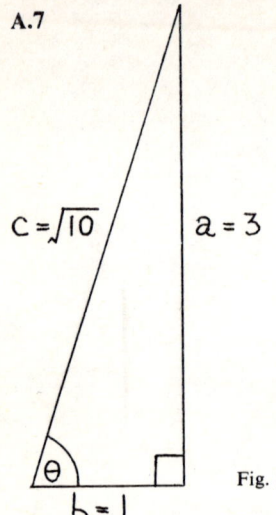

Fig. 3.43

$$2 \sin^2 \theta + 3 \cos^2 \theta = 2\tfrac{1}{10}$$

$$\tan \theta = \frac{\text{OPP}}{\text{ADJ}} = \frac{3}{1}$$

$$\text{Side } c = \sqrt{3^2 + 1^2}$$

$$= \sqrt{9 + 1}$$

$$\text{Side } c = \sqrt{10} = 3 \cdot 162$$

$$2 \sin^2 \theta + 3 \cos^2 \theta = 2\tfrac{1}{10}$$

Substituting

$$= 2 \times \tfrac{3^2}{10} + 3 \times \tfrac{1^2}{10}$$

$$= 2 \times \tfrac{9}{10} + \tfrac{3}{10}$$

$$= \tfrac{18}{10} + \tfrac{3}{10}$$

$$= \tfrac{21}{10} = 2\tfrac{1}{10} \quad Answer$$

A.8

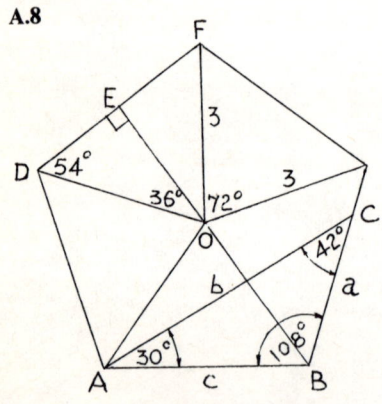

Fig. 3.44

$$\frac{DE}{OD} = \sin 36°$$

$$\therefore \quad DE = OD . \sin 36°$$

$$= 3 \times 0 \cdot 5878$$

$$DE = 1 \cdot 7634$$

$$\text{and} \qquad \frac{2}{DF = 3 \cdot 5268}$$

$$\frac{a}{\sin A} = \frac{c}{\sin C} \quad \therefore \quad \frac{a}{\sin 30°} = \frac{3 \cdot 5268}{\sin 42°}$$

$$a = \frac{3 \cdot 5268 . \sin 30°}{\sin 42°} = \frac{3 \cdot 5268 \times 0 \cdot 5}{0 \cdot 6691}$$

$$\therefore \quad a = 2 \cdot 63548 \text{ units}$$

$$b^2 = a^2 + c^2 - 2ac . \cos B$$

$$b^2 = 2 \cdot 63548^2 + 3 \cdot 5268^2 - 2 \times 2 \cdot 63548 \times 3 \cdot 5268 \times \cos 108°$$

$$b^2 = 6 \cdot 9455 + 12 \cdot 44 - 5 \cdot 27096 \times 3 \cdot 5268 \times -\cos 72°$$

$$b^2 = 19 \cdot 3855 + 18 \cdot 5898 \times 0 \cdot 3090$$

$$b^2 = 19 \cdot 3855 + 5 \cdot 7442$$

$$b = \sqrt{25 \cdot 1297} = 5 \cdot 013 \text{ units}$$

$$\text{Area of triangle } ABC = \sqrt{s(s-a)(s-b)(s-c)}$$

Where
$$s = \frac{a+b+c}{2} = \frac{2\cdot63548 + 5\cdot0130 + 3\cdot5268}{2}$$

$$\therefore \quad s = \frac{11\cdot17528}{2} = 5\cdot58764 \; units$$

$$(s-a) = 5\cdot58764 - 2\cdot63548 = 2\cdot95216 \;\Bigg]$$
$$(s-b) = 5\cdot58764 - 5\cdot0130 \;\; = 0\cdot57464 \;\Bigg\}$$
$$(s-c) = 5\cdot58764 - 3\cdot5268 \;\; = 2\cdot06084 \;\Bigg]$$

Substituting these values in the formula

$$Area \; \triangle ABC = \sqrt{5\cdot58764 \times 2\cdot95216 \times 0\cdot57464 \times 2\cdot06084}$$
$$= \sqrt{19\cdot53} = 4\cdot419 \; units^2$$

$$Area \; of \; pentagon = 5[\tfrac{1}{2}.(OD)(OF).\sin 72°]$$
$$= 5[\tfrac{1}{2} \times 3^2 \times 0\cdot9511]$$
$$= 5[4\cdot5 \times 0\cdot9511]$$
$$= 5 \times 4\cdot28 = 21\cdot40 \; units^2$$

$$\therefore \quad Ratio = \frac{21\cdot40}{4\cdot419} = 4\cdot85:1 \quad Answer$$

A.9

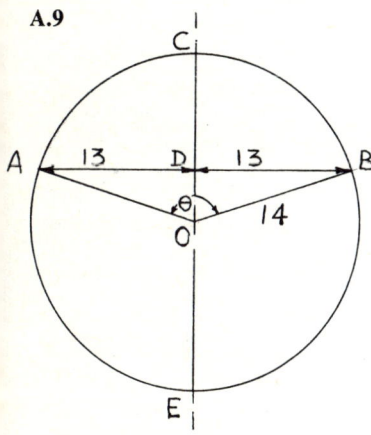

Fig. 3.45

$$OD = \sqrt{14^2 - 13^2}$$
$$= \sqrt{196 - 169}$$
$$= \sqrt{27}$$
$$\therefore \quad OD = 5\cdot196mm$$

$$\angle AOD = \sin \theta = \tfrac{13}{14} = 0\cdot9285 = 68°12'$$
$$\therefore \quad \angle \theta = \angle AOD \times 2 = 68°12' \times 2 = 136°24'$$

Length of Arc AEB = Circumference $-$ Arc ACB

$$= \pi D - \pi D \times \frac{\theta}{360°}$$

$$= \pi \times 28 - \pi \times 28 \times \frac{136°24'}{360°}$$

$$= \frac{22}{\underset{1}{7}} \times \frac{\overset{4}{28}}{1} - \frac{22}{\underset{1}{7}} \times \frac{\overset{4}{28}}{1} \times 0\cdot3788$$

$$= 88 - 33\cdot334$$

$$Arc \; AEB = 54\cdot666mm$$

$$Answer \qquad OD = 5\cdot196mm \;\Bigg]$$
$$\angle AOB = 136°24' \;\Bigg\}$$
$$Arc \; AEB = 54\cdot666mm \;\Bigg]$$

TEST PAPER 10

Q.1 Simplify:

(a)
$$\left(\frac{m^{\frac{3}{2}}n^{\frac{7}{4}}}{m^{\frac{3}{4}}n^{\frac{1}{2}}}\right)^{-\frac{1}{4}} \times [(\sqrt[3]{m^{-2}})(\sqrt[6]{n^{-1}})]^{-2}$$

(b)
$$\frac{10\frac{5}{16} - 5\frac{1}{4}}{5\frac{2}{5} \times 1\frac{1}{4}} \div \frac{24 - 21\frac{3}{4}}{4 \times 2\frac{1}{4}}$$

Q.2 Find the value of x by using log tables:

(a)
$$x = \frac{(84\cdot26^3 - 78\cdot4^2)^3}{111\cdot7^4}$$

(b)   When $R = e^{x+y}$ and $S = e^{x-y}$ find the value of (a) RS^2; (b) RS; (c) $\dfrac{R^2}{S}$;

Q.3 Find the value of x in

(a)
$$MO.\left(V - \frac{W}{X}\right) = G - K$$

(b)
$$Z.\left(X - \frac{Q}{X}\right) = KG$$

Q.4 Find the value of x in the following equations:

(a) By factorization $9x^2 - 5ax - bx = 0$

(b) By completing the square $3x^2 + 15x = 42$

Q.5 Plot the graph of $xy = 10$ for values of x between 2 and 8, then calculate the volume generated by rotating the graph about the x-axis.

Q.6 A rectangular room has a floor $ABCD$ and ceiling $A_1B_1C_1D_1$ where side $AB = $ 7m and $BC = AA_1 = 3$m. Find the length of side AC_1 (diagonal) and $\angle C_1AB_1$.

Q.7 A metal bar is stretched by $0\cdot15\%$ of its length on each loading. How many loadings will it take to increase its length by 3% of its original length?

Q.8 The segment of a circle has a diameter of 5cm and a depth of 1·5cm. Calculate (a) the area of the circle and (b) the area of the larger segment.

Q.9 In triangle XYZ; $\angle X = 90°$; $XY = 7$cm; $XZ = 6$cm. A line is drawn from point L which is 2cm from X on XY, to a point K on XZ; $\angle LKX = \angle XYZ$. Find the length of LK and the area of triangle LKX.

A.1 (a) $\left(\dfrac{m^{\frac{3}{4}}n^{\frac{7}{4}}}{m^{\frac{3}{8}}n^{\frac{1}{2}}}\right)^{-\frac{1}{3}} \times [(\sqrt[3]{m^{-2}})(\sqrt[6]{n^{-1}})]^{-2} = \left(\dfrac{m^{\frac{9}{8}}n^{\frac{7}{4}}}{m^{\frac{4}{8}}n^{\frac{2}{4}}}\right)^{-\frac{1}{3}} \times (m^{-\frac{2}{3}}n^{-\frac{1}{6}})^{-2}$

$$= (m^{\frac{1}{2}}n^{\frac{3}{4}})^{-\frac{1}{3}} \times (m^{\frac{4}{3}}n^{\frac{1}{3}})$$

$$= m^{-\frac{1}{6}}n^{-\frac{1}{4}} \times m^{\frac{4}{3}}n^{\frac{1}{3}}$$

$$= m^{\frac{-1+8}{6}} n^{\frac{-3+4}{12}}$$

$$= m^{\frac{7}{6}}n^{\frac{1}{12}} \quad Answer$$

(b)
$$\frac{10\frac{5}{16} - 5\frac{1}{4}}{5\frac{2}{5} \times 1\frac{1}{4}} \div \frac{24 - 21\frac{3}{4}}{4 \times 2\frac{1}{4}} = \frac{\dfrac{165}{16} - \dfrac{84}{16}}{\underset{1}{\dfrac{\overset{}{27}}{\cancel{5}}} \times \dfrac{\cancel{5}}{4}} \div \frac{\dfrac{96}{4} - \dfrac{87}{4}}{\underset{1}{\dfrac{4}{1}} \times \dfrac{9}{4}}$$

$$= \frac{\dfrac{81}{16}}{\dfrac{27}{4}} \div \frac{\dfrac{9}{4}}{\dfrac{1}{1}}$$

$$= \frac{\overset{3}{\cancel{81}}}{\underset{4}{\cancel{16}}} \times \frac{4}{\underset{1}{\cancel{27}}} \times \frac{4}{9} \times \frac{9}{1}$$

$$= 3 \quad Answer$$

A.2 (a)
$$x = \frac{(\overset{①}{84.26^3} - \overset{②}{78.4^2})^3}{\underset{③}{111.7^4}}$$

① $\quad 3 \log 84.26 = 3 \times 1.9256$ ② $\quad 2 \log 78.4 = 2 \times 1.8943$

$\qquad\qquad\quad = 5.7768 \qquad\qquad\qquad\qquad\quad = 3.7886$

$\qquad$ Antilogging $= \underline{598\,100}$ $\qquad$ Antilogging $= \underline{6\,146}$

$\qquad$ Subtracting ② from ① $\quad 598\,100$

$\qquad\qquad\qquad\qquad\qquad\qquad\quad \underline{6\,146}$

$\qquad\qquad\qquad\qquad\qquad\qquad\quad \overline{591\,954}$

$3 \log 591\,954 = \quad 5.7723$ $\qquad$ ③ $\quad 4 \times \log 111.7 = 4 \times 2.0479$

$\qquad\qquad\qquad\qquad \underline{\quad 3}$ $\qquad\qquad\qquad\qquad\qquad = 8.1916$

$\qquad\qquad\qquad\quad \overline{17.3169}$

$\qquad$ Subtract ③ $\qquad \underline{8.1916}$

$\qquad\qquad\qquad\qquad \overline{9.1253}$

$\qquad$ Antilogging $= 1335 \times 10^6 \quad Answer$

(b)
$$R = e^{x+y} \quad \text{and} \quad S = e^{x-y}$$

(i)
$$RS^2 = e^{x+y} \times e^{(x-y)2}$$
$$= e^{x+y+2x-2y} = e^{3x-y} \quad \text{Answer (i)}$$

(ii)
$$RS = e^{x+y+x-y} \quad = e^{2x} \quad \text{Answer (ii)}$$

(iii)
$$\frac{R^2}{S} = \frac{e^{(x+y)2}}{e^{(x-y)}} = e^{2x+2y-x+y}$$
$$= e^{x+3y} \quad \text{Answer (iii)}$$

A.3 (a)
$$MO.\left(V - \frac{W}{X}\right) = G - K$$
$$\left(V - \frac{W}{X}\right) = \frac{G-K}{MO}$$
$$\frac{W}{X} = V - \left(\frac{G-K}{MO}\right)$$
$$\therefore \quad X = \frac{W}{V - \left(\dfrac{G-K}{MO}\right)} \quad \text{Answer}$$

(b)
$$\therefore \quad \left(X - \frac{Q}{X}\right) = \frac{KG}{Z}$$
$$X = \frac{KG}{Z} + \frac{Q}{X}$$
$$X^2 = \frac{XKG}{Z} + Q$$
$$X^2 - \frac{XKG}{Z} - Q = 0$$

$$\left.\begin{array}{l} a = 1 \\ b = -\dfrac{KG}{Z} \\ c = -Q \end{array}\right\} \quad X = \frac{\dfrac{KG}{Z} \pm \sqrt{\left(-\dfrac{KG}{Z}\right)^2 - (4 \times 1 \times -Q)}}{2 \times 1}$$

$$X = \frac{\dfrac{KG}{Z} \pm \sqrt{\dfrac{(KG)^2}{Z^2} + 4Q}}{2}$$

$$\therefore \quad X = \frac{KG}{2Z} \pm \sqrt{\frac{K^2G^2}{Z^2} + 4Q} \quad \text{Answer}$$

A.4 (a)
$$9x^2 - 5ax - bx = 0$$
$$x(9x - 5a - b) = 0$$
$$\therefore \quad x = 0 \quad \text{OR} \quad 9x = 5a + b$$
$$\therefore \quad x = \frac{5a + b}{9}$$

Answer $\quad x = 0 \quad$ OR $\quad \dfrac{5a + b}{9}$

(b)
$$3x^2 + 15x = 42$$
$$x^2 + 5x = 14$$

When completing the square it is necessary to reduce the coefficient of x^2 to unity (1).

$$x^2 + 5x + (\tfrac{5}{2})^2 = 14 + (\tfrac{5}{2})^2$$
$$(x + \tfrac{5}{2})^2 = 14 + \tfrac{25}{4}$$
$$(x + \tfrac{5}{2}) = \pm\sqrt{\frac{56 + 25}{4}}$$
$$(x + \tfrac{5}{2}) = \pm\sqrt{\frac{81}{4}}$$
$$x = -\tfrac{5}{2} \pm \tfrac{9}{2}$$
$$\therefore \quad x = \tfrac{4}{2} = 2 \quad \text{OR} \quad x = -\tfrac{14}{2} = -7$$

Answer $\quad x = 2 \quad$ OR $\quad -7$

A.5
$$xy = 10 \quad \therefore \quad y = \frac{10}{x}$$

		a		b		c		d
x		2	3	4	5	6	7	8
$y = \dfrac{10}{x}$		5	$3\frac{1}{3}$	$2\frac{1}{2}$	2	$1\frac{2}{3}$	$1\frac{3}{7}$	$1\frac{1}{4}$

Using Simpson's 2nd Rule.

Ordinates	Radius	Areas πr^2	Simpson's Multipliers	Products
a	5	$\pi . 25$	1	$\pi . 25$
b	$2\frac{1}{2}$	$\pi . 6{\cdot}25$	3	$\pi . 18{\cdot}75$
c	$1\frac{2}{3}$	$\pi . 2{\cdot}78$	3	$\pi . 8{\cdot}34$
d	$1\frac{1}{4}$	$\pi . 1{\cdot}56$	1	$\pi . 1{\cdot}56$
				$\pi . 53{\cdot}65$

$$\tfrac{3}{8}h . (1 + 3 + 3 + 1) = \tfrac{3}{8} \times \tfrac{2}{1} \times 3{\cdot}1416 \times 53{\cdot}65$$
$$= \tfrac{3}{4} \times 168{\cdot}547$$
$$= \frac{505{\cdot}641}{4}$$
$$= 126{\cdot}41 \; units^3$$

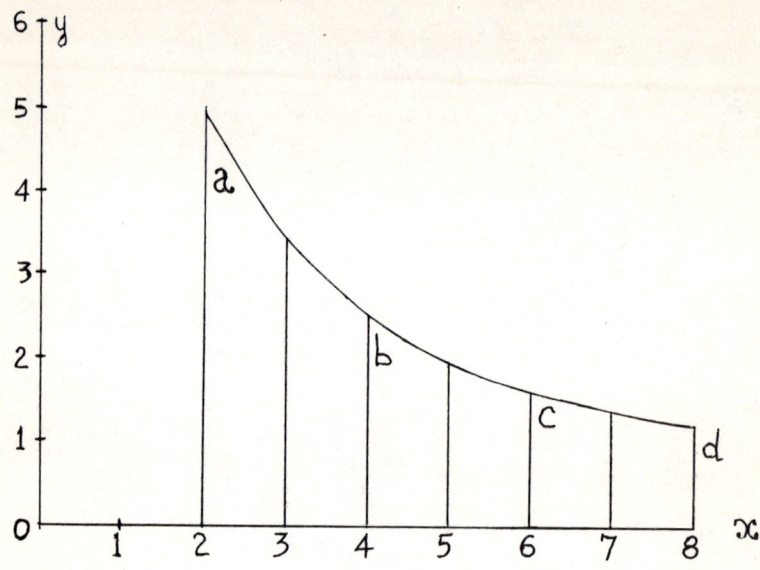

Fig. 3.46

Answer Volume = 126·41 units³

A.6

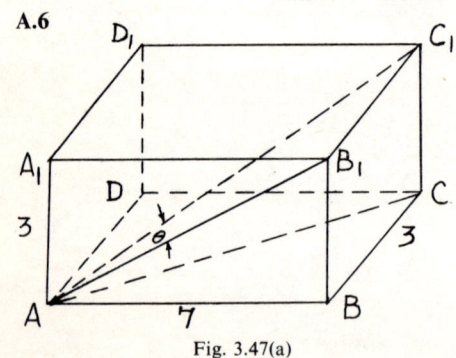

Fig. 3.47(a)

A_1B and $AC = \sqrt{7^2 + 3^2}$
$= \sqrt{49 + 9}$
$= \sqrt{58}$
$= 7.616m$

$$\therefore \quad AC_1 = \sqrt{7.616^2 + 3^2} = \sqrt{58 + 9} = \sqrt{67}$$
$$AC_1 = 8.185m$$

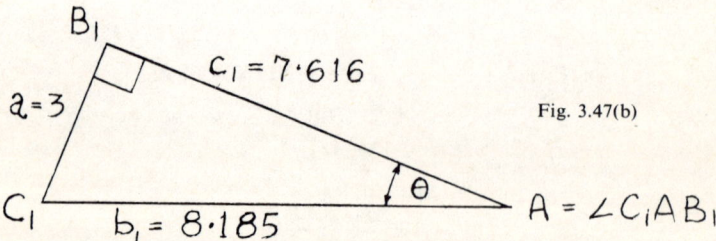

Fig. 3.47(b)

$$\frac{a}{b_1} = \sin \theta = \frac{3}{8 \cdot 185} = 0 \cdot 3665$$

$$\therefore \quad \angle C_1 A B_1 = 21° 30'$$

Answer   Diagonal $AC_1 = 8 \cdot 185\text{m}$

$$\left. \angle C_1 A B_1 = 21° 30' \right\}$$

A.7 $n\text{th term} = ar^{n-1} = 1 \cdot 03$

$$r = 1 \cdot 0015$$

$$\therefore \quad 1 \cdot 03 = ar^{n-1} = 1 \times 1 \cdot 0015^{n-1}$$

$$\log 1 \cdot 03 = (n - 1) \cdot \log 1 \cdot 0015$$

$$0 \cdot 0128 = (n - 1) \times 0 \cdot 0006$$

$$(n - 1) = \frac{0 \cdot 0128}{0 \cdot 0006} = 21$$

Answer   Number of loadings $= 21$

A.8 (a)   To find r for circle

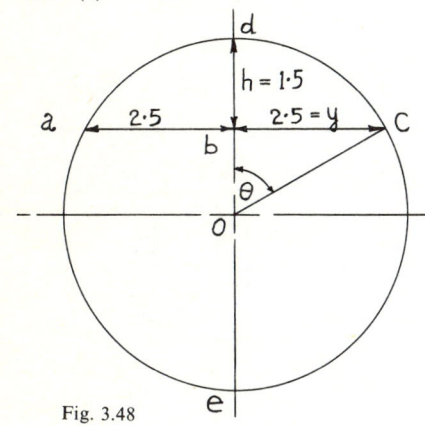

Fig. 3.48

$$r = \frac{y^2 + h^2}{2h} = \frac{2 \cdot 5^2 + 1 \cdot 5^2}{2 \times 1 \cdot 5}$$

$$r = \frac{6 \cdot 25 + 2 \cdot 25}{3} = \frac{8 \cdot 50}{3}$$

$$\therefore \quad r = 2 \cdot 833cm$$

$$\sin \theta = \frac{bc}{oc} = \frac{2 \cdot 5}{2 \cdot 833} = 0 \cdot 8825 = \quad 61°57'$$
$$\phantom{\sin \theta = \frac{bc}{oc} = \frac{2 \cdot 5}{2 \cdot 833} = 0 \cdot 8825} \quad\quad\quad 2$$
$$\angle aoc = \overline{123°54'}$$

$$Area\ of\ circle = \pi r^2 = \pi \times 2 \cdot 833^2$$

$$= \pi \times 8 \cdot 026$$

$$= 25 \cdot 2145cm^2 \quad (a)$$

(b)   *Area of small segment* $= \dfrac{\pi r^2 \theta}{360°} - \dfrac{r^2 \sin \theta}{2}$

$$= \left(\pi \times 2 \cdot 833^2 \times \frac{123°54'}{360°} \right) - \left(\frac{2 \cdot 833^2 \times \sin 123°54'}{2} \right)$$

$$= 25 \cdot 2145 \times 0 \cdot 3442 - \frac{8 \cdot 026 \times 0 \cdot 4150}{2}$$

$$= 8 \cdot 6788 - 3 \cdot 3308 = 5 \cdot 348cm^2$$

Area of large segment $= 25 \cdot 2145 - 5 \cdot 348$

$$= 19 \cdot 8665cm^2 \quad (b)$$

Answer   (a) $25 \cdot 2145cm^2$
(b) $19 \cdot 8665cm^2$

A.9

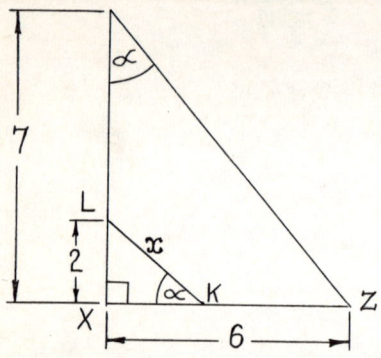

Fig. 3.49

By similar triangles

$$\frac{LX}{XZ} : \frac{LK}{YZ}$$

$$\frac{\overset{1}{2}}{\underset{3}{6}} = \frac{LK}{9\cdot22}$$

$$\text{Where } YZ = \sqrt{7^2 + 6^2} = \sqrt{85} = 9\cdot22 cm$$

$$\therefore \quad LK = \frac{9\cdot22}{3} = 3\cdot07 cm$$

$$\frac{XK}{YZ} : \frac{XL}{XZ} \qquad \therefore \quad \frac{XK}{7} = \frac{\overset{1}{2}}{\underset{3}{6}}$$

$$XK = \tfrac{7}{3} = 2\tfrac{1}{3} cm$$

$$\text{Area of } \triangle LKX = \frac{\overset{1}{2} \times 2\tfrac{1}{3}}{\underset{1}{2}} = 2\tfrac{1}{3} cm^2$$

Answer $\left. \begin{array}{l} LK = 3\cdot07 cm \\ \text{Area } \triangle LKX = 2\tfrac{1}{3} cm^2 \end{array} \right\}$

TEST PAPER 11

Q.1 Simplify

(a)
$$\left(\frac{(a^4)^{\frac{3}{2}}}{a^3} \right)^{\frac{1}{2}} \times a^{\frac{3}{4}}$$

(b)
$$\left(\frac{(a^{-\frac{4}{3}})^{\frac{1}{4}} \times a^{\frac{3}{4}} b^{\frac{3}{4}}}{b^{\frac{1}{4}} \cdot \sqrt[6]{b^3}} \right)^5$$

Q.2 By the use of logarithms find the value of a and n in the following equations

(i) $387 = a3^n$
(ii) $2000 = a7\cdot9^n$

Q.3 Make A the subject in the following expression

$$V = \sqrt{\dfrac{2gh}{1 + \dfrac{4fla^2}{DA^2}}}$$

Q.4 In the following expression, solve for y when $x = \frac{3}{2}$

$$y = ax + \frac{b}{x^2}$$

When $y = 5; x = \frac{1}{2}$ and when $y = 3; x = 1$.

Q.5 Determine graphically the values of x and y which satisfy the following equations

(i) $xy = 18$
(ii) $x + 2y = 12$

Q.6 Given a quadrilateral $ABCD$, $ABC = 90°$, $AB = 4$m, $BC = 3$m, $CD = 4\cdot4$m, and $DA = 7\cdot4$m. Find the remaining angles.

Q.7 Prove that

$$\tan \frac{A}{2} = \frac{\sin A}{1 + \cos A}$$

Q.8 A $2\cdot5$cm diameter hole is drilled through a cone from base to apex. The cone base diameter is 6cm and the vertical height is 9cm. Find how much material is removed.

Q.9 Two cords AB and CD intersect at O touching the circumference of a circle at points $ABCD$. Given that $AO = AC$ prove that $OD = DB$.

A.1 (a)
$$\left(\frac{(a^4)^{\frac{3}{2}}}{a^3}\right)^{\frac{1}{2}} \times a^{\frac{3}{2}} = \left(\frac{a^{10}}{a^3}\right)^{\frac{1}{2}} \times a^{\frac{3}{2}}$$

$$= \frac{a^5}{a^{\frac{3}{2}}} \times a^{\frac{3}{2}}$$

$$= a^{\frac{10+5-3}{2}}$$

$$= a^{\frac{12}{2}} = a^6 \quad \textit{Answer}$$

(b)
$$\left(\frac{(a^{-\frac{4}{5}})^{\frac{1}{2}} \times a^{\frac{3}{5}}b^{\frac{3}{6}}}{b^{\frac{1}{2}} \cdot \sqrt[6]{b^3}}\right)^5 = \left(\frac{a^{-\frac{2}{5}} \times a^{\frac{3}{5}}b^{\frac{1}{2}}}{b^{\frac{1}{2}} \cdot b^{\frac{1}{2}}}\right)^5$$

$$= \left(\frac{a^{\frac{3-2}{5}}}{b^{\frac{6-3}{6}}}\right)^5$$

$$= \left(\frac{a^{\frac{1}{5}}}{b^{\frac{1}{2}}}\right)^5$$

$$= \frac{a}{b^{\frac{5}{2}}} \quad \textit{Answer}$$

A.2

    (i) $387 = a3^n$

    (ii) $2000 = a7 \cdot 9^n$

    (ii) $\log 2000 = n \log 7 \cdot 9 + \log a$

    (i) $\log 387 = n \log 3 + \log a$

    (ii) $3 \cdot 3010 = n \times 0 \cdot 8976 + \log a$

    (i) $2 \cdot 5877 = n \times 0 \cdot 4771 + \log a$

    $\overline{0 \cdot 7133 = n \times 0 \cdot 4205}$

$$\therefore \quad n = \frac{0 \cdot 7133}{0 \cdot 4205} = 1 \cdot 696$$

Substitute $n = 1 \cdot 696$ in (ii) above

    (ii) $3 \cdot 3010 = 1 \cdot 696 \times 0 \cdot 8976 + \log a$

$$\therefore \quad \log a = 3 \cdot 3010 - 1 \cdot 5220$$

$$\therefore \quad \log a = 1 \cdot 7790$$

$$a = 60 \cdot 12$$

$$Answer \quad \left. \begin{array}{l} n = 1 \cdot 696 \\ a = 60 \cdot 12 \end{array} \right\}$$

A.3

$$V = \sqrt{\frac{2gh}{1 + \dfrac{4fla^2}{DA^2}}}$$

$$V^2 = \frac{2gh}{1 + \dfrac{4fla^2}{DA^2}}$$

$$1 + \frac{4fla^2}{DA^2} = \frac{2gh}{V^2}$$

$$\frac{4fla^2}{DA^2} = \frac{2gh}{V^2} - 1$$

$$4fla^2 = \left(\frac{2gh}{V^2} - 1\right) . DA^2$$

$$A^2 = \frac{4fla^2}{\left(\dfrac{2gh}{V^2} - 1\right) . D}$$

$$A = \sqrt{\frac{4fla^2}{\left(\dfrac{2gh}{V^2} - 1\right) . D}}$$

$$A = 2a \sqrt{\frac{fl}{\left(\dfrac{2gh}{V^2} - 1\right) . D}} \quad Answer$$

A.4

$$y = ax + \frac{b}{x^2}$$

When $y = 5$; $x = \frac{1}{2}$ (i)

and

when $y = 3$; $x = 1$ (ii)

To find b and substituting values, we get

(i) $5 = \frac{1}{2}a + 4b$
(ii) $3 = a + b$

Multiplying (i) by 2

(i) $10 = a + 8b$
(ii) $\underline{3 = a + b}$
$\ 7 = 7b$

$\therefore \quad b = \frac{7}{7} = 1$

Substituting $b = 1$ in (ii) above

$$3 = a + b$$

$$\therefore \quad 3 = a + 1$$

$$\text{and} \quad a = 3 - 1 = 2$$

To find y when $x = \frac{3}{2}$

$$y = ax + \frac{b}{x^2}$$

$$\therefore \quad y = (2 \times \tfrac{3}{2}) + \frac{1}{(\frac{3}{2})^2}$$

$$y = 3 + \tfrac{4}{9} = 3\tfrac{4}{9} \text{ when } x = \tfrac{3}{2} \quad \textit{Answer}$$

A.5   (i) $xy = 18$
(ii) $x + 2y = 12$

From (i) $\qquad y = \dfrac{18}{x}$

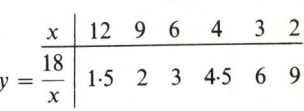

x	12	9	6	4	3	2
$y = \dfrac{18}{x}$	1·5	2	3	4·5	6	9

From (ii) $y = \dfrac{12 - x}{2} = 6 - \dfrac{x}{2} = \dfrac{-x}{2} + 6$

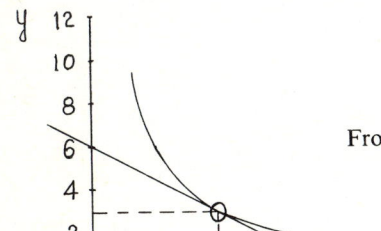

Fig. 3.50

x	2	6	10
$-\dfrac{x}{2}$	-1	-3	-5
$+6$	$+6$	$+6$	$+6$
y	5	3	1

$x = 6$; $y = 3$ *Answer*

A.6

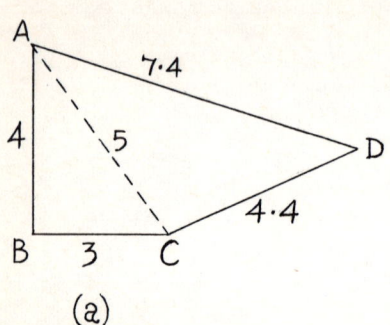

(a)

(b)

Fig. 3.51

By Pythagoras

$$(AC)^2 = (AB)^2 + (BC)^2$$
$$= 4^2 + 3^2$$
$$= 16 + 9$$
$$AC = \sqrt{25} = 5$$

$\angle A$ in $\triangle ABC$

$$= \tfrac{3}{4} = \tan A = 0.75 = 36°52'$$

$$\cos A = \frac{b^2 + c^2 - a^2}{2bc}$$

$$= \frac{7.4^2 + 5^2 - 4.4^2}{2 \times 7.4 \times 5}$$

$$\cos A = \frac{54.76 + 25 - 19.36}{74} = \frac{60.40}{74}$$

$$\cos A = 0.8162 \qquad \therefore \quad \angle A = 35°18'$$

$\angle A$ in quadrilateral $\qquad 36°52' + 35°18'$

$$\angle A = 72°10'$$

$$\cos C = \frac{a^2 + b^2 - c^2}{2ab} = \frac{4.4^2 + 7.4^2 - 5^2}{2 \times 4.4 \times 7.4}$$

$$= \frac{19.36 + 54.76 - 25}{65.12}$$

$$= \frac{49.12}{65.12} = 0.7543$$

$$\therefore \quad \angle C \ (\angle D \ in \ quad) = 41°02'$$

$\angle C$ in quadrilateral $= 360° - (72°10' + 90° + 41°02')$

$$360° - 203°12' = 156°48'$$

$$\therefore \quad \left.\begin{array}{l} \angle A = 72°10' \\ \angle C = 156°48' \\ \angle D = 41°02' \end{array}\right\} \ Answer$$

A.7

$$\tan \frac{A}{2} = \frac{\sin A}{1 + \cos A}$$

LHS

$$= \frac{\sin \dfrac{A}{2}}{\cos \dfrac{A}{2}}$$

126

$$= \sqrt{\frac{\left|\dfrac{1 - \cos A}{2}\right|}{\left|\dfrac{1 + \cos A}{2}\right|}}$$

$$= \sqrt{\frac{1 - \cos A}{2} \times \frac{\cancel{2}}{1 + \cos A}}$$

Multiplying by

$$\frac{1 + \cos A}{1 + \cos A} = \sqrt{\frac{(1 - \cos A)(1 + \cos A) \leftarrow (a^2 - b^2)}{(1 + \cos A)(1 + \cos A) \leftarrow (a + b)^2}}$$

$$= \sqrt{\frac{(1 - \cos^2 A)}{(1 + \cos A)^2}}$$

$$= \sqrt{\frac{\sin^2 A}{(1 + \cos A)^2}}$$

$$= \frac{\sin A}{1 + \cos A} = \text{RHS} \quad \textit{Answer}$$

A.8

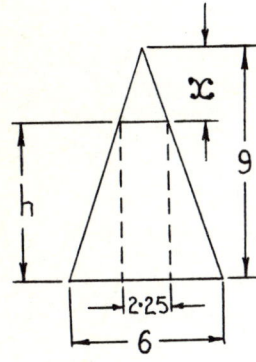

Fig. 3.52

By similar triangles

$$\frac{6}{9} = \frac{2 \cdot 25}{x}$$

$$\therefore \quad x = \frac{2 \cdot 25 \times 9}{6} = 3 \cdot 375 cm$$

$$\therefore \quad h = 9 - 3 \cdot 375 = 5 \cdot 625 cm$$

Volume of cylinder removed

$$= \frac{\pi}{4} d^2 h = \frac{\pi}{4} \times 2 \cdot 25^2 \times 5 \cdot 625$$

$$= \frac{\pi}{4} \times 5 \cdot 0625 \times 5 \cdot 625 = 22 \cdot 37 cm^3$$

Volume of cone removed

$$= \frac{\pi}{12} d^2 h \quad \text{where } h = x \text{ above}$$

$$= \frac{\pi}{12} \times 2 \cdot 25^2 \times 3 \cdot 375$$

$$= 0 \cdot 33 \times 0 \cdot 7854 \times 5 \cdot 0625 \times 3 \cdot 375$$

$$= 4 \cdot 47 cm^3$$

Total Volume Removed

$$= 22{\cdot}37 + 4{\cdot}47$$

$$= 26{\cdot}84\text{cm}^3 \quad Answer$$

A.9

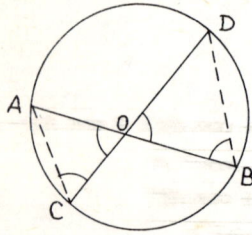

Given: Two chords AB and CD,

$\quad AO = AC$, circle with points at A, B, C and D.

To Prove: $OD = DB$

Construction: None

Fig. 3.53

Proof: In $\triangle$'s AOC and ODB

$$A\hat{C}O = C\hat{O}A \quad \text{Isosceles triangle}$$

$$D\hat{O}B = C\hat{O}A \quad \text{Opposite angles}$$

$$\angle C = \angle B \quad \text{On arc } AD$$

$$\therefore \quad \triangle DOB \text{ is isosceles}$$

then $OD = DB \quad Answer$

TEST PAPER 12

Q.1 Simplify

(a)
$$\sqrt[3]{\frac{a^{\frac{1}{2}}b^{\frac{1}{4}}a^{-\frac{2}{3}}}{a^{-3}b^5}}$$

(b)
$$\frac{a^{-5}(a^6 + 7a^5 + 6a^4)}{(a + 6)^2}$$

Q.2 Evaluate

$$\frac{\left(\dfrac{2{\cdot}568}{0{\cdot}825}\right)^{2{\cdot}56} \times 12}{\log_e \dfrac{0{\cdot}825}{2{\cdot}256} \times 4{\cdot}578^{2{\cdot}46}}$$

Q.3 If $a = 1 - \dfrac{1}{b}$ and $b = 1 + \dfrac{2}{c}$ find an expression for c and prove that

$$a + b + \frac{1}{c} = \frac{(2 - a)(1 + 2a)}{2(1 - a)}$$

128

Q.4 Factorize the following

(a)
$$15 - \frac{3}{x} + \frac{5}{x^2} - \frac{1}{x^3}$$

(b)
$$6r^2 s^2 - 35qrs + 11q^2$$

(c)
$$6p^2 qr - 4pq^2 r - 9p^2 qs + 6pq^2 s$$

(d)
$$a^2 c - a^2 d - b^2 c + b^2 d$$

Q.5 Draw the graph of $\dfrac{x^2}{4} + 2$ between -2 and $+5$. From this solve $x^2 + 2x = 4$, deducing the straight line graph to solve this.

Q.6 In any triangle ABC and using the usual sign convention for a triangle prove that
$$\cot A - \cot B = \frac{b^2 - a^2}{ab \sin C}$$

Q.7 A cone measures 3m at the base with a vertical height of 4·5m. When the cone is resting on its base the vertical height of the water in the cone is 2m. Find the height of the water if the cone was lying with its apex vertically downwards.

Q.8 A conical vessel with a height to diameter ratio of 4:6 has 16cm of water in it. 200 metal spheres of diameter 1·5cm are dropped in; find the increase in the height of the water in the cone.

Q.9 In any triangle ABC (not a right-angled triangle) if D is the mid-point of BC, prove that
$$(AB)^2 + (AC)^2 = 2(AD)^2 + 2(BD)^2$$

A.1 (a)
$$\frac{\sqrt[3]{a^{\frac{2}{3}} b^{\frac{1}{2}} a^{-\frac{2}{3}}}}{a^{-3} b^5} = \left(\frac{a^{\frac{3-9+6}{2}}}{b^{\frac{10-1}{2}}} \right)^{\frac{1}{3}}$$
$$= \frac{a^{\frac{1}{3}}}{b^{\frac{3}{2}}}$$
$$= \frac{\sqrt[3]{a}}{\sqrt{b^3}} \quad Answer$$

(b)
$$\frac{a^{-5}(a^6 + 7a^5 + 6a^4)}{(a + 6)^2} = \frac{a^{-5} \times a^4 (a^2 + 7a + 6)}{(a + 6)(a + 6)}$$
$$= \frac{(a + 1)(a + 6)}{a(a + 6)(a + 6)}$$
$$= \frac{(a + 1)}{a(a + 6)} \quad Answer$$

A.2

$$\frac{\left(\dfrac{2\cdot568}{0\cdot825}\right)^{2\cdot56} \times 12}{\log_e \dfrac{0\cdot825}{2\cdot256} \times 4\cdot578^{2\cdot46}} = \frac{\log_e \dfrac{2\cdot256}{0\cdot825} \times \left(\dfrac{2\cdot568}{0\cdot825}\right)^{2\cdot56} \times 12}{4\cdot578^{2\cdot46}}$$

$\log 2\cdot568 = 0\cdot4096$	$\log_e 8\cdot25 = 2\cdot1102$
$\log 0\cdot825 = \overline{1}\cdot9165$	$\log_e 10^{-1} = \overline{3}\cdot6974$
$\overline{0\cdot4931}$	$\log_e 0\cdot825 = \overline{1}\cdot8076$
$2\cdot56 \times 0\cdot4931$	
$= 1\cdot2623$	$\log_e 2\cdot256 = 0\cdot8136$
	$\log_e 0\cdot825 = \overline{1}\cdot8076$
$\log\left(\dfrac{2\cdot568}{0\cdot825}\right)^{2\cdot56} = 1\cdot2623$	$\log_e \dfrac{2\cdot256}{0\cdot825} = 1\cdot0060$
$\log 1\cdot0060 = 0\cdot0025$	
$\log 12 = 1\cdot0792$	$2\cdot46 \times \log 4\cdot578$
$\log \text{topline} = \overline{2\cdot3440}$	$= 2\cdot46 \times 0\cdot6607$
$\log 4\cdot578^{2\cdot46} = 1\cdot6253$	$= 1\cdot6253$
$\overline{0\cdot7187}$	

Antilogging $= 5\cdot232$ *Answer*

A.3

$$a = 1 - \frac{1}{b}; \qquad b = 1 + \frac{2}{c}$$

also

$$\frac{1}{b} = 1 - a \qquad \therefore \quad b = \frac{1}{1-a}$$

and

$$\frac{2}{c} = b - 1 \qquad \therefore \quad c = \frac{2}{b-1}$$

Substituting for c in terms of a

$$c = \frac{2}{b-1} = \frac{2}{\dfrac{1}{1-a}-1} = \frac{2(1-a)}{1-(1-a)} = \frac{2(1-a)}{1-1+a} = \frac{2(1-a)}{a}$$

and

$$\frac{1}{c} = \frac{a}{2(1-a)}$$

Substituting in

$$a + b + \frac{1}{c} = a + \frac{1}{1 - a} + \frac{a}{2(1 - a)}$$

$$= \frac{2a(1 - a) + 2 + a}{2(1 - a)}$$

$$= \frac{2a - 2a^2 + 2 + a}{2(1 - a)} = \frac{-2a^2 + 3a + 2}{2(1 - a)}$$

Factorizing $= \dfrac{-2a^2 - a + 4a + 2}{2(1 - a)}$

$$= \frac{-a(2a + 1) + 2(2a + 1)}{2(1 - a)} = \frac{(2 - a)(2a + 1)}{2(1 - a)}$$

$$= \text{RHS} \quad \textit{Answer}$$

A.4 **(a)**

$$15 - \frac{3}{x} + \frac{5}{x^2} - \frac{1}{x^3} = 15 + \frac{5}{x^2} - \frac{3}{x} - \frac{1}{x^3}$$

$$= 5\left(3 + \frac{1}{x^2}\right) - \frac{1}{x}\left(3 + \frac{1}{x^2}\right)$$

$$= \left(5 - \frac{1}{x}\right)\left(3 + \frac{1}{x^2}\right) \quad \textit{Answer}$$

(b) $6r^2s^2 - 35qrs + 11q^2 = 6r^2s^2 - 2qrs - 33qrs + 11q^2$ $\quad\quad P = 66q^2r^2s^2$

$$= 2rs(3rs - q) - 11q(3rs - q) \quad\quad\quad S = -35qrs$$

$$= (2rs - 11q)(3rs - q) \quad \textit{Answer}$$

(c) $6p^2qr - 4pq^2r - 9p^2qs + 6pq^2s = 2pqr(3p - 2q) - 3pqs(3p - 2q)$

$$= (2pqr - 3pqs)(3p - 2q)$$

$$= (3p - 2q)(2r - 3s)(pq) \quad \textit{Answer}$$

(d) $a^2c - a^2d - b^2c + b^2d = a^2(c - d) - b^2(c - d)$

$$= (a^2 - b^2)(c - d)$$

$$= (a + b)(a - b)(c - d) \quad \textit{Answer}$$

A.5
$$\frac{x^2}{4} + 2 = y$$

x	-2	-1	0	1	2	3	4	5
x^2	4	1	0	1	4	9	16	25
$\dfrac{x^2}{4}$	1	$\frac{1}{4}$	0	$\frac{1}{4}$	1	$2\frac{1}{4}$	4	$6\frac{1}{4}$
$+2$	2	2	2	2	2	2	2	2
y	3	$2\frac{1}{4}$	2	$2\frac{1}{4}$	3	$4\frac{1}{4}$	6	$8\frac{1}{4}$

$$x^2 + 2x = 4 \qquad \therefore \quad x^2 + 2x - 4 = 0$$

Divide through by 4; $= \dfrac{x^2}{4} + \dfrac{x}{2} - 1 = 0$

$$\left.\begin{array}{l} \dfrac{x^2}{4} + 2 \quad\quad = y \\[2mm] \dfrac{x^2}{4} + \dfrac{x}{2} - 1 = 0 \end{array}\right\} \text{ subtracting}$$

$$-\frac{x}{2} + 3 \quad\quad = y$$

x	-2	$+2$	$+5$
$-\dfrac{x}{2}$	1	-1	$-2\frac{1}{2}$
$+3$	3	3	3
y	4	2	$\frac{1}{2}$

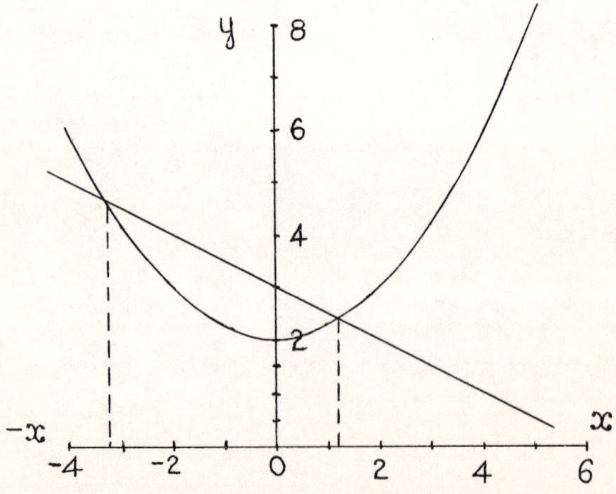

Fig. 3.54

From graph $x = -3 \cdot 2$ OR $1 \cdot 25$ *Answer*

A.6 From Fig. 3.55

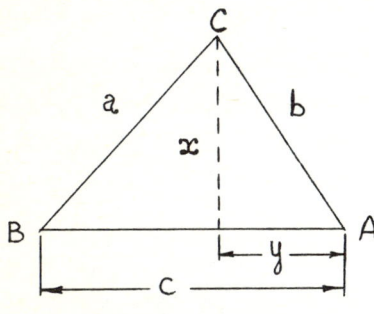

Fig. 3.55

$$\cot A - \cot B = \frac{y}{x} - \frac{(c - y)}{x}$$

$$= \frac{y - c + y}{x} = \frac{2y - c}{x} \quad \text{(i)}$$

and

$$\frac{x}{a} = \sin B$$

From sine rule:

$$\frac{a}{\sin A} = \frac{b}{\sin B} = \frac{c}{\sin C}$$

$$\sin B = \frac{b \cdot \sin C}{C}$$

$$\therefore \quad \frac{x}{a} = \frac{b \cdot \sin C}{c} \quad \text{and} \quad x = \frac{ba \cdot \sin C}{c}$$

Substituting for x in (i) above

$$\cot A - \cot B = \frac{(2y - c)c}{ba \cdot \sin C} = \frac{2yc - c^2}{ba \cdot \sin C} \quad \text{(iv)}$$

From figure and using Pythagoras

$$(c - y)^2 = a^2 - x^2 \quad \text{(ii)}$$

also

$$y^2 = b^2 - x^2 \quad \text{(iii)}$$

From (ii)

$$c^2 - 2cy + y^2 = a^2 - x^2$$

$$\therefore \quad 2cy = c^2 + y^2 - a^2 + x^2$$

Substitute for $2yc$ in (iv) above

$$\cot A - \cot B = \frac{\cancel{c^2} + y^2 - a^2 + x^2 - \cancel{c^2}}{ba \cdot \sin C}$$

and substituting for y^2 in (iii) above

$$\cot A - \cot B = \frac{b^2 - \cancel{x^2} - a^2 + \cancel{x^2}}{ba \cdot \sin C}$$

$$\therefore \quad \cot A - \cot B = \frac{b^2 - a^2}{ab \cdot \sin C} \quad \textit{Answer}$$

A.7

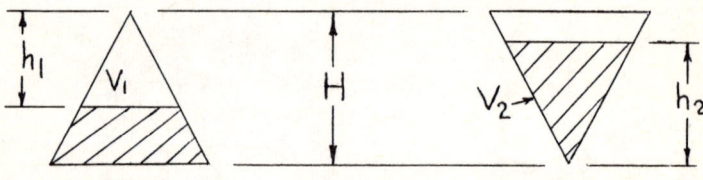

Fig. 3.56

$$\frac{V}{v_1} = \frac{H^3}{h_1^3}$$

$$\frac{V}{v_1} = \frac{4 \cdot 5^3}{2 \cdot 5^3} = \frac{91 \cdot 125}{15 \cdot 625}$$

$$\frac{V}{v_1} = 5 \cdot 85$$

$$\therefore \quad V = 5 \cdot 85 v_1$$

and

$$v_1 = 0 \cdot 171 V$$

$$\frac{V}{v_2} = \frac{V}{0 \cdot 829 V} = \frac{4 \cdot 5^3}{h_2^3}$$

$$\therefore \quad h_2^3 = \frac{4 \cdot 5^3 \times 0 \cdot 829 \cancel{V}}{\cancel{V}}$$

$$h_2 = 4 \cdot 5 \times \sqrt[3]{0 \cdot 829}$$

$$= 4 \cdot 5 \times 0 \cdot 94$$

$$\therefore \quad h_2 = 4 \cdot 23 \text{m} \quad \textit{Answer}$$

A.8

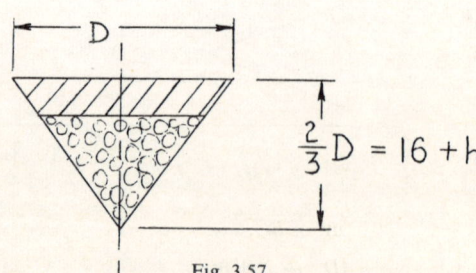

$$\frac{2}{3}D = 16 + h$$

Fig. 3.57

$$\text{Volume of cone} = \frac{\pi D^2}{12} \cdot h \quad \text{OR} \quad \tfrac{1}{3}\pi r^2 \cdot h$$

$$\text{Volume of spheres} = \tfrac{4}{3}\pi r^3$$

If $h = \tfrac{2}{3}D = 16$ then $D = 24$cm

$$\textit{Volume of water} = \tfrac{1}{3}\pi r^2 h$$

$$= \tfrac{1}{3}\pi \times 12^2 \times 16$$

$$= 3{\cdot}1416 \times 768 = \textit{2413}{\cdot}\textit{4cm}^3$$

$$\textit{Volume of metal spheres} = 200 \times \tfrac{4}{3}\pi r^3 = 200 \times \tfrac{4}{3} \times 3{\cdot}1416 \times 0{\cdot}75^3$$

$$= 4{\cdot}189 \times 0{\cdot}421 \times 200 = \textit{352}{\cdot}\textit{7cm}^3$$

$$\textit{Total volume} \text{ (water + spheres)} = 2413{\cdot}4 + 352{\cdot}7 = \textit{2766}{\cdot}\textit{1cm}^3$$

Increase in height of water

$$\pi D^2 h = 2766$$

Substituting $\tfrac{2}{3}D$ for h

$$\frac{\pi D^2 \cdot \tfrac{2}{3}D}{12} = 2766$$

$$\frac{2\pi D^3}{3} \times \tfrac{1}{12} = 2766$$

$$D^3 = \frac{\overset{18}{\cancel{36}} \times 2766}{2\pi}$$

$$D = \sqrt[3]{\frac{18 \times 2766}{3{\cdot}1416}}$$

$$D = \sqrt[3]{15\,842} = \textit{25}{\cdot}\textit{11cm}$$

$$h = \tfrac{2}{3}D = \tfrac{2}{3} \times 25{\cdot}11 = \textit{16}{\cdot}\textit{74cm}$$

Increase in height $= 16{\cdot}74 - 16 = 0{\cdot}74$cm   *Answer*

A.9

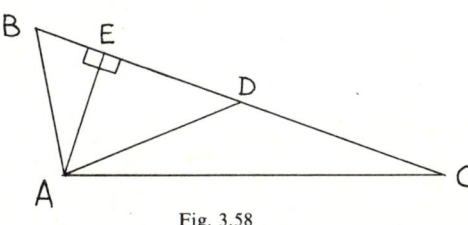

Fig. 3.58

Given: $\triangle ABC$ and D as mid-point of BC.

To Prove: $(AB)^2 + (AC)^2 = 2(AD)^2 + 2(BD)^2$.

Construction: From BD construct perpendicular to point A.

Proof:

and
$$(AB)^2 = (AE)^2 + (BE)^2 \left.\right\}$$
$$(AC)^2 = (AE)^2 + (CE)^2 \left.\right\} \text{ Adding}$$
$$(AB)^2 + (AC)^2 = 2(AE)^2 + (BE)^2 + (CE)^2$$

But
$$(AE)^2 = (AD)^2 - (DE)^2$$
$$\therefore \quad (AB)^2 + (AC)^2 = 2[(AD)^2 - (DE)^2] + (BE)^2 + (CE)^2$$
$$= 2(AD)^2 - 2(DE)^2 + (BE)^2 + (CE)^2$$
$$= 2(AD)^2 + (BE)^2 - (DE)^2 + (CE)^2 - (DE)^2$$

Rearranging
$$= 2(AD)^2 + (BE + DE)(BE - DE) + (CE + DE)(CE - DE)$$
$$= 2(AD)^2 + (BD)(BE - DE) + (CE + DE)(BD)$$
$$= 2(AD)^2 + (BD)(BE) - \cancel{(BD)(DE)} + (BD).(CE) + \cancel{(BD)(DE)}$$
$$= 2(AD)^2 + (BD)[(BE) + (CE)]$$
$$= 2(AD)^2 + (BD)(BC)$$
$$= 2(AD)^2 + (BD)(2BD)$$
$$\therefore \quad (AB)^2 + (AC)^2 = 2(AD)^2 + 2(BD)^2 \quad \textit{Answer}$$